Ornamental Palms

Production and Processing

The Authors

Dr. P. Ranchana is working as a Post Doctoral Fellow in Tamil Nadu Agricultural University, Coimbatore. She has published 6 international, 5 national and 30 popular articles, 6 manuals and 5 book chapters. She served as a Course Associate for UG and PG courses with respect to Floriculture and Landscaping. She is a life member of several professional societies. She handled classes for ODL students towards Floriculture and Landscaping aspects. She assisted in the conduct of 25 days training programmes for women entrepreneurs at the Department of Floriculture and Landscaping, TNAU, Coimbatore.

Dr. M. Kannan is presently working as the Professor and Head, Department of Floriculture and Landscaping, Horticultural College and Research Institute, Tamil Nadu Agricultural University, Coimbatore. He started his carrier in 1986 as Assistant Professor (Hort.) at Tamil Nadu Agricultural University, Yercaud. To his credit he has published more than 9 books, several technical bulletins, student manuals and book chapters related to Horticulture and specifically to floriculture. During the tenure of his service spanning over 31 years, Dr. Kannan has been involved in UG/PG/Ph.D teaching, research on floriculture and many extension oriented activities. He is the recipient of 10 awards and is a member of various professional and academic societies. He was the Project Officer for the Tamil Nadu Precision Farming Project which sensitized several horticultural farmers of Tamil Nadu on hi-tech production of horticultural crops using drip and fertigation.

Ornamental Palms

Production and Processing

P. Ranchana

M. Kannan

Department of Floriculture and Landscaping
Tamil Nadu Agricultural University, Coimbatore

2018

Daya Publishing House®

A Division of

Astral International Pvt. Ltd.

New Delhi – 110 002

Cataloging in Publication Data--DK
Courtesy: D.K. Agencies (P) Ltd. <docinfo@dkagencies.com>

Ranchana, P., author.
Ornamental palms : production and processing / P. Ranchana, M. Kannan.
   pages cm
   ISBN 9789387057265 (Interational Edition)

      1. Palms. 2. Plants, Ornamental.   I. Kannan, M. (College teacher of floriculture and landscaping), author. II. Title.

   SB413.P17R36 2017   DDC 635.977484   23

Published by   : **Daya Publishing House®**
                                    A Division of
                                    Astral International Pvt. Ltd.
                                    –ISO 9001:2015 Certified Company –
                                    4736/23, Ansari Road, Darya Ganj
                                    New Delhi-110 002
                                    Ph. 011-43549197, 23278134
                                    E-mail: info@astralint.com
                                    Website: www.astralint.com

Preface

Palms are the first modern family of monocots that is clearly represented in the fossil record and first appeared in the fossil record around 80 million years ago. The first modern species of palm appeared 65 million years ago, confirmed by fossil records. They appear to have undergone an early period of extreme adaptation. By 60 million years ago, many of the modern, specialized genera of palms appeared and became widespread.

They are one of the most well known and extensively cultivated plant families. They have had an important role to humans throughout much of history. Ornamental palms are widely used in landscaping for their exotic appearance, making them one of the most economically important plants in the world. In many historical cultures, palms are symbols for such ideas as victory, peace, and fertility. Today, palms remain a popular symbol for the tropics and vacation spots.

This book gives a clear picture about different types of palms based on nature of root, trunk, leaves, growth, flowers and fruits. Moreover, it act as a quick reference to know about the palms suitable for different regions (tropical, sub tropical, temperate), dry, wet soils and coastal regions, as house plants, in landscaping as hedges and screens. This book is bestowed with the description of 101 palms with coloured photographs which are widely used in different parts of the world as ornamental, medicinal plant and also for culinary purposes. Colourful photographs on value added products of coconut were taken from a Coir fair held at Coimbatore.

This book will be highly helpful as a basic guide to know about palms for the students, budding landscapists and landscaping consultants. It will create curiosity among the plant lovers to include palm as one of the important ornamental plant component in landscaping of both urban and rural areas. At this juncture, we would like to thank all the staff and students of Department of Floriculture and Landscaping, Horticultural College and Research Institute, TNAU, Coimbatore who extended co-operation during preparation of this book. We express our deep sense of gratitude to Daya Publishing House, a Division of Astral International Pvt. Ltd., New Delhi for taking up this assignment within a short span of time.

P. Ranchana

M. Kannan

Contents

1

Introduction

Palms are monocots, included in the section of Angiosperms characterized by bearing a single seed leaf. Scientifically, palms are classified as belonging to the family Palmae, are perennial and distinguished by having woody stems. The palm family consists of six subfamilies, each representing a major line of evolution. The Coryphoideae is the subfamily with the most unspecialized characters. It is followed by the Calamoideae, Nypoideae, Ceroxyloideae, Arecoideae and Phytelephantoideae subfamilies; the last exhibiting the greatest number of specialized characters. All palms are part of the largest family of flowering plants belonging to the monocot division, as like Bermuda grass, corn, wheat, onion and banana.

Geographically, palms can be found in habitats ranging from southern France where the European fan palm (*Chamaerops humilis*) naturally occurs at 44° north latitude, to Chatham Island, New Zealand, at 44° south latitude, where the shaving brush palm (*Rhopalostylis sapida*) is native. However, despite this impressive spread of latitude, the overwhelming majority of palm species are native to the tropical regions of the earth. There are about 130 palm species occur naturally beyond the tropical latitudes (23.5° N. & S.).

Palms prefer to grow in sparse to dense clusters, although several palm genera have both solitary and clustering members. They grow in a variety of sizes. The Sea coconut is a palm with the largest seeds of any plant of about 20 inches in diameter and weighing over 132 kg each. Raphia palms have leaves up to 75 feet long and 15 feet wide and are known to have the largest leaves of any palms. The *Corypha umbraculifera* (Talipot Palm) palms have the largest flower of any palms, up to 25 feet tall and containing millions of small flowers. Colombia's national tree,

Ceroxylon quindiuense (Wase Palm), is the tallest palm tree in the world, reaching a height of about 180 feet tall. The fruit of most palms are somewhat unusual in shape and have tendency to float. Seed distribution in the coconut palm is exceptional as the coconut, when mature, falls onto the beach and is carried on the tide, sometimes for long distances, to germinate elsewhere. Floating coconuts are common and capable of germinating after floating in the sea for up to 110 days.

Raphia palm leaves (75 feet long and 15 feet wide)

***Corypha* palm flower (up to 25 feet tall and containing millions of small flowers)**

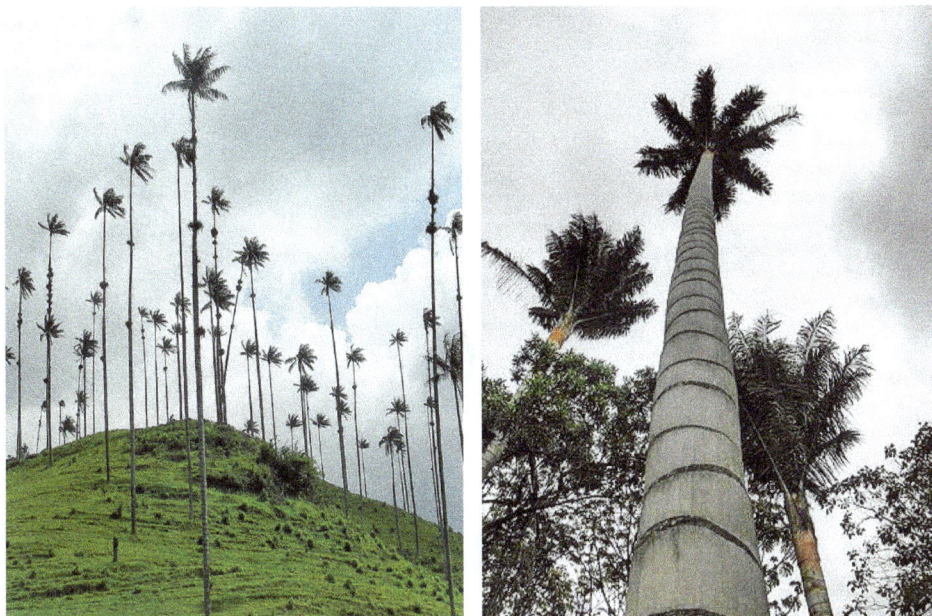

***Ceroxylon quindiuense*, the tallest palm tree in the world (reaching heights of 180 feet tall)**

The production of ornamental palms is relatively an emerging industry in floriculture sector and getting popular now-a-days as house plants, filler material in bouquet making, flower arrangement and also used in landscaping (as specimen plant for hedging and screening purpose and to give elegant look in front of the building) for their exotic appearance, making them one of the most economically important plants in the world. In many historical cultures, palms are used as sign of victory, peace, and fertility. Today, palms remain a popular symbol for the tropics and vacation spots. There are roughly 2600 currently known species of ornamental palms, most of which are restricted to the tropical, subtropical and the warmer climates of the world.

2

Classification of Palms

a) The nature of roots

The roots of palms are usually tough and thick, although slender and wiry types are available. Some roots creep over the soil's surface, while a few palms even develop aerial roots that may eventually make them appear to be growing on stilts.

A few palms develop aerial roots at or near the base of the trunk or stems. These include *Cryosophila warscewiczii* (Rootspine palm) and *Clinostigma exorrhizum*. Other palms with aerial roots include *Pinanga aristata, Verschaffeltia splendida* (Seychelles stilt palm) and *Socratea exorrhiza*. Some of these develop into stilt or prop roots, which often form a cone and give the palm extra support, they usually occurs on plams native to wet soils and humid conditions, whereas the base of the palms is at risk from decay. Indeed, in a few instances, the lower part of the trunk withers and the palm is entirely supported by these roots.

Aerial roots on the trunk
(*Cryosophila warscewiczii*)

Aerial roots at the base of the trunk
(*Clinostigma exorrhizum*)

b) The nature of trunks

Trunks vary in diameter, as a result of later upward growth of the palm. Most trunks are straight, but a few are swollen, perhaps the best known of these being *Hyophorbe lagenicaulis* (Bottle plam) and *Hyophorbe verschaffeltii* (Spindle palm). The trunk of *Hyophorbe lagenicaulis* has a bulbous base, while that of *Hyophorbe verschaffeltii* is narrow at its crown and base, but widens at its centre.

Trunk surface also varies, whereas some are relatively smooth, such as *Jubaea chilensis* (Chilean wine palm, Coquito palm), the trunk of *Aiphanes aculeata* (Ruffle palm, Chonta Ruro) is ringed with black spines. Some palms have a mass of fibres on their trunks, such as *Trachycarpus fortunei* (Chinese windmill palm, Chusan palm).

The range of trunks includes some with distinctive growth rings (formed of scars from old leaves) such as *Howea belmoreana* (Belmore sentry palm, Curly palm), *Adonidia merrillii* (Christmas palm, Manila palm) has indistinct growth rings, while *Phoenix roebelenii* (Dwarf date plam, Roebelin palm) has dominant, persistent leaf bases.

Hyophorbe lagenicaulis - Bottle palm

Hyophorbe verschaffeltii - Spindle palm

Jubaea chilensis – Smooth trunk

Aiphanes aculeata - Trunk ringed with black spines

Trachycarpus fortunei – Trunk with mass of fibres

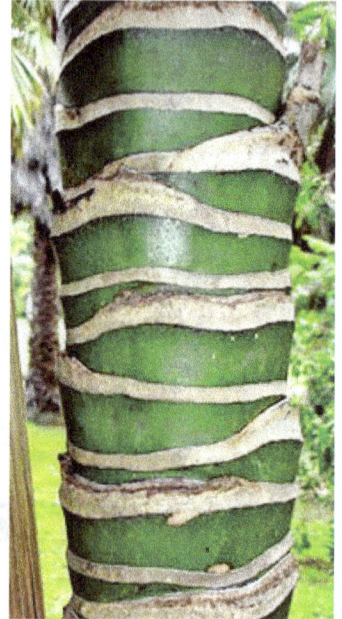

Howea belmoreana – Trunk with distinctive growth rings

Adonidia merrillii - Trunk with indistinctive growth rings

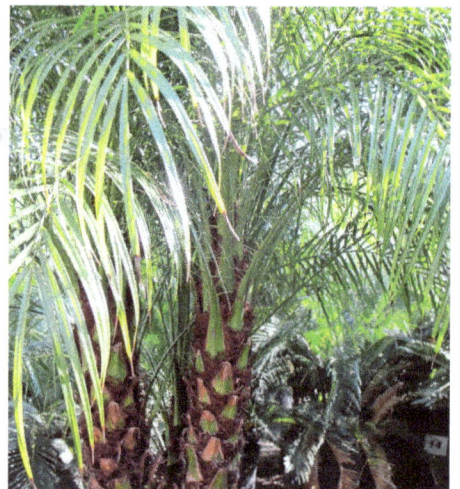

Phoenix roebelenii - Trunk with persistent leaf bases

c) The nature of leaves

Also known as fronds, the leaves of palms have varied natures, some are borne at the top of a palm in a distinctive crown, while others have leaves on stems that develop from bottom portion of the trunk or even from stems that arise directly from the ground.

Fan-leaved palms

Also known as Fan palms, these have leaves with circular or semi-circular outline. The leaves resemble fans and are either partly divided, known as palmafid or referred to as palmate if totally divided. The leaf divisions are known as segments.

Palms with fan -shaped leaves include *Chamaerops humilis* (European Fan Palm, Mediterranean Fan Palm), *Trachycarpus fortunei* (Chinese Windmill Palm, Chusan Palm) and *Washingtonia filifera* (Californian Fan Palm, Petticoat palm). *Licuala spinosa* (Mangrove Fan Palm, Spicy Licuala) has wedge shaped leaflets.

Feather-type palms

Also known as Pinnate palms, these palms have fronds that are divided on either side of the mid-rib, with the impression of many leaflets. They are sometimes described as the backbones and ribs of a fish.

Palms that reveal this style include *Howea belmoreana* (Belmore Sentry Palm or Curly Palm), *Howea forsteriana* (Kentia Palm or Thatch Leaf Palm) and *Lytocaryum weddelianum* (Dwarf Coconut Palm or Weddel Palm).

The leaflets also known as pinnae, are variable in size, shape and the angle at which they are attached to the central leaf-stem or rachis. Some arise from the rachis in a flat, even formation, such as *Archontophoenix alexandrae* (Alex Palm, Northern Bangalow Palm) and *Archontophoenix cunninghamiana* (Piccabeen Bangalow Palm). Some, such as *Howea belmoreana* (Belmore Sentry Palm or Curly Palm) create an angled, upright, perhaps V-shaped formation, while others droop. Some have leaflets in two distinct planes, such as *Phoenix canariensis* (Canary Island Date Palm), while others have an uneven distribution along the rachis, including the beautiful and popular *Hydriastele wendlandiana* (Florence Falls Palm, Latrum Palm).

Pinnate leaves exhibit an extreme size-range in the Palmae, varying from (including the petiole) 1 m in length in species of *Chamaedorea* to 25 m long in *Raphia regalis*. The latter is reputed to be a world record for the plant kingdom. All five major economic palms have pinnate leaves: coconut (*Cocos nucifera*), African oil palm *(Elaeis guineensis)*, date palm (*Phoenix dactylifera*), betel nut palm (*Areca catechu*) and peach palm (*Bactris gasipaes*).

A further type has a plumose (feather-like) and irregular arrangement of leaflets, such as *Syagrus romanzoffiana* (Giriba Palm, Queen Palm).

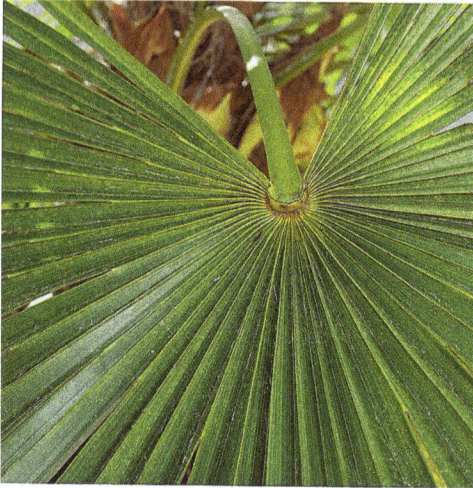

Trachycarpus fortunei – Fan-leaved palm

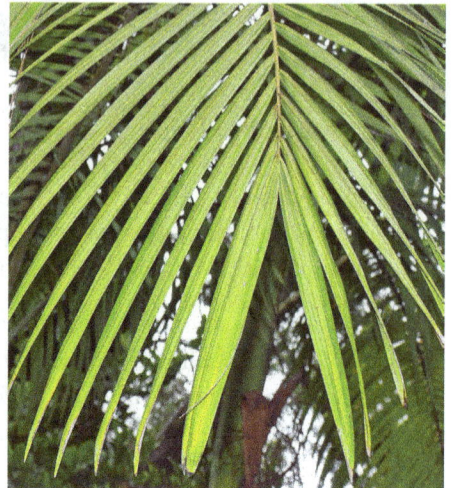

Archontophoenix alexandrae - Feather-leaved palm

Fishtail palms

Also known as bipinnate palms, these have leaves that are twice divided and have a fishtail appearance at their tips. Examples are *Caryota mitis* (Clustered Fishtail Palm, Tufted Fishtail Palm) and *Caryota urens* (Fishtail Palm, Toddy palm).

Caryota mitis – Fishtail palm

Entire-leaved palms

These attractive palms are unusal in that leaves are undivided in their mature state. One example of this type is *Johannesteijsmannia altifrons* (Diamond Joey).

Johannesteijsmannia altifrons **(Entire-leaved palm)**

d) The nature of growth

Palms have a wide range of growth habits, from those that are trunkless to those with a clumping nature or majestic palms with tall trunks. Some even have a climbing nature.

Trunkless palms

Several palms appear to be trunkless, including those that later form dominant trunks but, while young, seem to have only a mass of leafstalks growing from bottom portion of the palm and even from the ground level, Often, palms native to scrubby land or dense florists' are trunkless.

A few palms have trunks that are wholly or partly beneath the soil's surface, including *Sabal minor* (Blue palmetto palm, Little blue stem) and *Serenoa repens* (Saw Palmetto, Scrub Palmetto).

Sabal minor - **Trunkless palm**

Clumping palms

These palms have multiple trunks or stems. They include popular species such as *Dypsis lutescens* (Golden cane palm, Yellow bamboo palm), *Phoenix dactylifera* (Date palm, Date) and *Phoenix reclinata* (African date palm, Senegal date palm).

Sometimes, a clump of palms is initially formed of many stems, but later one or two dominant. An example for this is *Chamaerops humilis*. When some clump-forming palms are grown as ornamental features, shoots at the base of the plant are removed to produce a single trunk with a dominant, uncluttered nature.

Dypsis lutescens -**Clumping palm**

Solitary palms

Palms with solitary trunks include *Adonidia merrillii* (Christmas palm, Manila palm), *Livistona australis* (Australian cabbage palm, Gippsland palm), *Phoenicophorium borsigianum* (Latanier palm), *Trachycarpus fortunei* (Chinese windmill palm, Chusan palm) and *Washingtonia filifera* (Californian fan palm, Petticoat palm).

Adonidia merrillii - **Solitary palm**

Branching palms

Few palms have a natural branching nature although branching can sometimes occur as the result of damage to the growing point of a trunk. *Hyphaena thebaica* (Gingerbread palm) is native to northern and northeast Africa, where it has a natural, freely branching and clustering nature. *Serenoa repens* (Saw Palmetto, Scrub Palmetto) is a trunkless palm with lateral underground stems that send up new stems.

Hyphaena thebaica - **Branching palm**

Climbing Palms

More than 600 palms have a climbing nature, including *Calamus australis* (Lawyer cane, Rattan palm), with its scrambling, climbing, clustering nature and stems that grows up to 25 m high in the wild area.

Many palms with a climbing nature initially form clumps, with stems usually known as canes, a few have a single climbing stem. Occasionally, the canes spread over the ground until they find a plant to which they can cling. Many climbing palms have canes clothed in spines and hooks. These enable to cling to a host and give it protection from animals. The clusters of flowers also aid in supporting canes, as many have spine-like hooks.

Calamus australis - **Climbing Palm**

e) The nature of flowers

The arrangement of flowers in palms varies widely, including those at the top of the stems that grow from ground level and those in the heads or crowns of palms.

Palm bears its flower when they mature and it varies from one type to another, Plams with a relatively low stature may flower earlier in their lives than tall types. For example *Chamaedorea elegans* (Good luck plam, Parlour palm) which grows up to 3 m high and is often grows as an indoor palm, may flower when four to six years old, whereas *Lodoicea maldivica,* which can be 25 m high or more in the wild areas, can be 40 or more years old when it bears flowers.

Some palms are monocarpic, indicating that a stem or trunk dies once it has borne flowers. Unless the palm has a clumping nature or is formed of many individual stems , it results in death of entire plant.

Several palms produce their flowers terminally and towards the top of the trunk, whereas others develop axillary inflorescences from shoots at the junctions of leaf stalks or just below the crownshaft of palms with pinnate (feather-like) leaves.

The individual flowers of palms usually small and are borne in clusters that sometimes bear more than a million individual flowers. Some are attractive and also fragrant, such as *Coccothrinax fragrans* (Fragrant Cuban thatch, Yuraguana) and *Hyophorbe verschaffelti* (Spindle palm) while others like *Arenga pinnata* (Sugar palm, Gomuti palm) have drooping clusters of purple flowers with a rather unpleasant aroma.

Some palms have flowers that have a bisexual nature, with both male and female parts present. They are called hermaphrodite. Others have a unisexual nature, with only one sex present. Unisexual flowers can either be borne separately on the same inflorescence or in a separate inflorescences on the same palm. Finally many species have male and female flowers on separate palms. Examples of rapid sexual maturity are found among *Chamaedorea* spp., whereas the buri palm (*Corypha utan*) shows late maturity.

Unlike many flowers on trees, those on palms are relatively short-lived and usually last for only about a day. They are either pollinated by insects, such as honeybees, flies and beetles or by the wind.

f) The nature of fruits

Palm fruits and seeds vary in size and colour. They are botanically either a drupe (fleshy and usually with a single stone-like seed, like a plum or cherry) or single-seeded berries. Many palm fruits are edible and succulent, perhaps the best known being *Phoenix dactylifera* (Date palm), which has been part of the diet of millions of people for centuries, especially in warm regions.

In terms of weight and size, palm seeds exhibit extreme differences. An individual seed of the popular ornamental parlor palm (*Chamaedorea elegans*) weighs only 0.23 g, as compared to the massive seed of the double coconut (*Lodoicea maldivica*) which weighs as much as 20 kg. The double coconut has the distinction of bearing the largest seed in the plant kingdom.

Chamaedorea elegans- seeds   *Lodoicea maldivica*- seed

3

Landscape Uses of Palms

Palms suitable as houseplants

1. *Aiphanes aculeata* (Chonta Ruro, Coyure Palm, Ruffle Palm, Spine Palm)

2. *Archontophoenix alexandrae* (Alex palm, Alexandra king palm, Alexander king palm, King palm, Northern bungalow palm)

3. *Archontophoenix cunninghamiana* (Bangalow palm, Piccabeen bungalow palm, Piccabean palm)

4. *Arenga caudata* (Dwarf sugar palm)

5. *Carpentaria acuminata* (Carpentaria palm)

6. *Caryota mitis* (Burmese fishtail palm, Clustered fishtail palm, Fishtil palm and Tufted fishtail palm)

7. *Caryota urens* (Fishtail palm, Jaggery palm, Kitul tree, Sago palm, Solitary fishtail palm, Toddy palm, Wine palm)

8. *Chamaedorea cataractarum* (Cascade palm, Cat palm, Cataract palm)

9. *Chamaedorea elegans* (Good luck palm, Parlor palm, Parlour palm)

10. *Chamaedorea erumpens* (Bamboo palm)

11. *Chamaedorea metallica* (Metallic palm, Miniature fishtail plam)

12. *Chamaedorea seifrizii* (Bamboo palm, Reed palm)

13. *Chamaedorea stolonifera*

14. *Chamaerops humilis* (European fan palm, Fan palm, Mediterranean fan palm)

15. *Dypsis decaryi* (Three-sided palm, Triangle palm)

16. *Dypsis lutescens* (Areca palm, Butterfly palm, Cane palm, Golden feather palm, Golden yellow palm, Madagascar palm, Yellow palm, Yellow butterfly palm)

17. *Hedyscepe canterburyana* (Big mountain palm, Umbrella palm)

18. *Howea belmoreana* (Belmore sentry palm, Curly palm, Sentry palm)

19. *Howea forsteriana* (Forster sentry palm, Kentia palm, Paradise palm, Sentry palm, Thatch leaf palm)

20. *Laccospadix australasica* (Atherton palm, Queensland kentia)

21. *Licuala grandis* (Fan-leaved palm, Palas payung, Ruffled fan palm, Vanuatu fan palm)

22. *Licuala orbicularis*

23. *Linospadix monostachya* (Walking stick palm)

24. *Livistona australis* (Australian cabbage palm, Australian palm, Australian fan palm, Cabbage palm, Fan palm and Gippsland palm)

25. *Livistona chinensis* (Chinese fan palm, Chinese fountain palm, Fan palm, Footstool palm, Fountain palm)

26. *Phoenix canariensis* (Canary island date, Canary island date palm, Canary date palm)

27. *Phoenix roebelenii* (Dwarf date palm, Miniature date palm, Pygmy date palm, Roebelin palm)

28. *Pinanga coronata* (Bunga, Ivory cane palm, Pinang palm)

29. *Ptychosperma elegans* (Alexander palm, Solitaire palm)

30. *Reinhardtia gracilis* (Window palm, Window pane palm)

31. *Rhapis excelsa* (Bamboo palm, Fern rhapis, Ground rattan, Lady palm, Little lady palm, Slender lady palm, Miniature fan palm)

32. *Rhapis humilis* (Reed rhapis, Slender lady palm)

33. *Syagrus romanzoffiana* (Giriba palm, Queen palm)

34. *Trachycarpus fortunei* (Chinese windmill palm, Chusan palm, Fan palm, Hemp palm, Windmill palm)

35. *Washingtonia filifera* (American cotton palm, Caliornian cotton palm, Californian fan palm, Caliornian palm, Cotton palm, Desert palm, Desert fan palm, Desert Washingtonia, Fan palm, Northern Washingtonia, Petticoat palm)

36. *Washingtonia robusta* (Mexican Washington palm, Mexican Washingtonia, Mexican fan palm, Skyduster Southern Washingtonia, Thread palm)

Palms as ground covers

1. *Arenga caudata* (Dwarf sugar palm)

2. *Chamaedorea cataractarum* (Cascade palm, Cat palm, Cataract palm)

3. *Chamaedorea elegans* (Good luck palm, Parlor palm, Parlour palm)

4. *Chamaedorea metallica* (Metallic palm, Miniature fishtail plam)

5. *Chamaedorea stolonifera*

6. *Reinhardtia gracilis* (Window palm, Window pane palm)

7. *Sabal minor* (Blue palmetto palm, Bush palmetto, Dwarf palm, Dwarf palmetto, Dwarf palmetto palm, Little blue stem, Scrub palmetto, Swamp palmetto)

Palms for hedging and screening purposes

1. *Acoelorraphe wrightii* (Everglades palm, Paurotis palm, Saw cabbage palm, Silver saw palm, Silver saw palmetto)

2. *Arenga caudata* (Dwarf sugar palm)

3. *Arenga engleri* (Dwarf sugar palm, Sugar palm, Formosa palm)

4. *Bactris gasipaes* (Chonta, Peach palm, Pejibaya, Pejibeye, Pejivalle, Pewa, Pupunha)

5. *Caryota mitis* (Burmese fishtail palm, Clustered fishtail palm, Fishtil palm and Tufted fishtail palm)

6. *Chamaedorea cataractarum* (Cascade palm, Cat palm, Cataract palm)

7. *Chamaedorea microspadix* (Bamboo palm, Hardy bamboo palm)

8. *Chamaedorea seifrizii* (Bamboo palm, Reed palm)

9. *Chamaerops humilis* (European fan palm, Fan palm, Mediterranean fan palm)

10. *Dypsis lutescens* (Areca palm, Butterfly palm, Cane palm, Golden feather palm, Golden yellow palm, Madagascar palm, Yellow palm, Yellow butterfly palm)

11. *Metroxylon sagu* (Sago palm)

12. *Oncopserma tigillarium* (Nibung palm)

13. *Phoenix reclinata* (African date palm, African wild date palm, Senegal date palm)

14. *Pinanga coronata* (Bunga, Ivory cane palm, Pinang palm)

15. *Ptychosperma macarthurii* (Hurricane palm, Macarthur feather palm, Macarthur palm)

16. *Rhapidophyllum hystrix* (Blue palmetto, Creeping palmetto, Dwarf saw palmetto, Hedgehog palm, Needle palm, Porcupine palm, Spine palm, Vegetable porcupine)

17. *Rhapis excelsa* (Bamboo palm, Fern rhapis, Ground rattan, Lady palm, Little lady palm, Slender lady palm, Miniature fan palm)

18. *Serenoa repens* (Saw palmetto, Scrub palmetto)

4

Palms for Different Locations

Palms for only tropical regions

1. *Cyrtostachys renda* (Sealing wax palm, Lipstick palm, Maharajah palm, Pinang-rajah)

2. *Licuala grandis* (Fan-leaved palm, Palas payung, Ruffled fan palm, Vanuatu fan palm)

Palms for both tropical and subtropical regions

1. *Aiphanes aculeata* (Chonta Ruro, Coyure Palm, Ruffle Palm, Spine Palm)

2. *Archontophoenix alexandrae* (Alex palm, Alexandra king palm, Alexander king palm, King palm, Northern bungalow palm)

3. *Caryota mitis* (Burmese fishtail palm, Clustered fishtail palm, Fishtil palm and Tufted fishtail palm)

4. *Caryota urens* (Fishtail palm, Jaggery palm, Kitul tree, Sago palm, Solitary fishtail palm, Toddy palm, Wine palm)

5. *Chamaedorea seifrizii* (Bamboo palm, Reed palm)

6. *Dictyosperma album* (Common princess palm, Hurricane palm, Princess palm)

7. *Hydriastele wendlandiana* (Florence falls palm, Latrum palm)

8. *Hyophorbe lagenicaulis* (Bottle palm)

9. *Hyophorbe verschaffeltii* (Spindle palm)

10. *Laccospadix australasica* (Atherton palm, Queensland kentia)

11. *Licuala orbicularis*

12. *Lytocaryum weddellianum* (Dwarf coconut palm, Sago palm, Weddel palm)

13. *Normanbya normanbyi* (Black palm)

14. *Pinanga coronata* (Bunga, Ivory cane palm, Pinang palm)

15. *Ptychosperma macarthurii* (Hurricane palm, Macarthur feather palm, Macarthur palm)

16. *Reinhardtia gracilis* (Window palm, Window pane palm)

17. *Roystonea regia* (Cuban royal , Cuban royal palm, Florida royal palm, Royal palm)

Palms for subtropical and warm - temperate regions

1. *Arenga engleri* (Dwarf sugar palm, Sugar palm, Formosa palm)

2. *Brahea armata* (Blue fan palm, Blue hesper palm, Grey goddess, Mexican blue fan palm, Mexican blue palm, Short blue hesper)

3. *Butia capitata* (Butia palm, Jelly palm, Pindo palm, South American jelly palm, Wine palm)

4. *Chamaedorea elegans* (Good luck palm, Parlor palm, Parlour palm)

5. *Chamaedorea metallica* (Metallic palm, Miniature fishtail plam)

6. *Chamaerops humilis* (European fan palm, Fan palm, Mediterranean fan palm)

7. *Hedyscepe canterburyana* (Big mountain palm, Umbrella palm)

8. *Howea belmoreana* (Belmore sentry palm, Curly palm, Sentry palm)

9. *Howea forsteriana* (Forster sentry palm, Kentia palm, Paradise palm, Sentry palm, Thatch leaf palm)

10. *Linospadix monostachya* (Walking stick palm)

11. *Livistona australis* (Australian cabbage palm, Australian palm, Australian fan palm, Cabbage palm, Fan palm and Gippsland palm)

12. *Rhapis excelsa* (Bamboo palm, Fern rhapis, Ground rattan, Lady palm, Little lady palm, Slender lady palm, Miniature fan palm)

13. *Rhapis humilis* (Reed rhapis, Slender lady palm)

14. *Trachycarpus fortunei* (Chinese windmill palm, Chusan palm, Fan palm, Hemp palm, Windmill palm)

Palms for coastal regions

1. *Copernicia macroglossa* (Cuban petticoat palm, Jata de Guanbacoa, Petticoat palm)

2. *Phoenix dactylifera* (Date, Date palm)

3. *Pseudophoenix sargentii* (Buccaneer palm, Cherry palm, Florida cherry palm)

4. *Serenoa repens* (Saw palmetto, Scrub palmetto)

5. *Thrinax radiata* (Florida thatch palm, Jamaica thatch, Sea thatch, Silk-top thatch)

Palms for dry soils

1. *Arenga pinnata* (Aren, Areng palm, Black fibre palm, Gomuti palm, Kabong, Sugar palm)

2. *Bismarckia nobilis* (Bismarck palm)

3. *Borassus flabellifer* (Doub palm, Lontar plam, Palmyra palm, Pannamaram, Tala palm, Talauriksha palm, Tal-gas, Toddy palm, Wine palm)

4. *Brahea armata* (Blue fan palm, Blue hesper palm, Grey goddess, Mexican blue fan palm, Mexican blue palm, Short blue hesper)

5. *Butia capitata* (Butia palm, Jelly palm, Pindo palm, South American jelly palm, Wine palm)

6. *Chamaerops humilis* (European fan palm, Fan palm, Mediterranean fan palm)

7. *Coccothrinax argentata* (Florida silver palm, Silver palm, Silver thatch palm, Silvertop, Silver top palm)

8. *Coccothrinax crinita* (Mat palm, Old man palm, Old man thatch palm, Palma petate, Thatch palm)

9. *Copernicia macroglossa* (Cuban petticoat palm, Jata de Guanbacoa, Petticoat palm)

10. *Jubaeopsis caffra* (Pondoland palm)

11. *Latania loddigesii* (Blue latan, Blue latan palm)

12. *Latania lontaroides* (Red latan, Red latan palm)

13. *Latania verschaffeltii* (Yellow, Yellow latan palm)

14. *Livistona chinensis* (Chinese fan palm, Chinese fountain palm, Fan palm, Footstool palm, Fountain palm)

15. *Phoenix reclinata* (African date palm, African wild date palm, Senegal date palm)

16. *Phoenix rupicola* (Cliff date palm, Date palm, East India wine palm, India date palm, Wild date palm)

17. *Pseudophoenix sargentii* (Buccaneer palm, Cherry palm, Florida cherry palm)

18. *Sabal minor* (Blue palmetto palm, Bush palmetto, Dwarf palm, Dwarf palmetto, Dwarf palmetto palm, Little blue stem, Scrub palmetto, Swamp palmetto)

19. *Sabal palmetto* (Blue palmetto, Cabbage palm, Cabbage palmetto, Cabbage tree, Common Palmetto, Palmetto, Palmetto palm)

20. *Serenoa repens* (Saw palmetto, Scrub palmetto)

21. *Thrinax parviflora* (Broom palm, Florida thatch palm, Iron thatch, Mountain thatch palm, Jamaica thatch palm, Palmetto thatch, Thatch, Thatch pole)

22. *Thrinax radiata* (Florida thatch palm, Jamaica thatch, Sea thatch, Silk-top tahtch)

23. *Trithrinax acanthacoma* (Buriti palm, Spiny fibre palm)

Palms for wet and swampy soils

1. *Acoelorraphe wrightii* (Everglades palm, Paurotis palm, Saw cabbage palm, Silver saw palm, Silver saw palmetto)

2. *Copernicia macroglossa* (Cuban petticoat palm, Jata de Guanbacoa, Petticoat palm)

3. *Cyrtostachys renda* (Sealing wax palm, Lipstick palm, Maharajah palm, Pinang-rajah)

4. *Elaeis guineensis* (African oil palm, Jacara palm)

5. *Euterpe edulis* (Assai palm, Jacara palm)

6. *Hydriastele wendlandiana* (Florence falls palm, Latrum palm)

7. *Livistona australis* (Australian cabbage palm, Australian palm, Australian fan palm, Cabbage palm, Fan palm and Gippsland palm)

8. *Metroxylon sagu* (Sago palm)

9. *Phoenix roebelenii* (Dwarf date palm, Miniature date palm, Pygmy date palm, Roebelin palm)

10. *Ptychosperma macarthurii* (Hurricane palm, Macarthur feather palm, Macarthur palm)

11. *Raphia farinifera* (Madagascar raffia palm, Latrum palm)

12. *Ravenea rivularis* (Majesty palm)

13. *Rhapidophyllum hystrix* (Blue palmetto, Creeping palmetto, Dwarf saw palmetto, Hedgehog palm, Needle palm, Porcupine palm, Spine palm, Vegetable porcupine)

14. *Roystonea regia* (Cuban royal , Cuban royal palm, Florida royal palm, Royal palm)

5

False Palms

The term "palm," correctly-applied, refers to plants which are members of the Palmae, but by popular usage has also been applied to plants which resemble palms in some ways. Seven plants have a common name which includes the word "palm," but they are not real palms in the scientific sense. They are:

Traveller's palm (*Ravenala madagascariensis*)

- It is a woody tree with a palm-like stem

- It belongs to the family Strelitziaceae

- It is native to Madagascar and widely cultivated as an ornamental throughout the tropics. Individual leaves bear greater resemblance to a banana plant (to which it is related) than a palm; they are arranged in two distinct ranks in the same plane and forms a fan-shaped head. Flowers of the traveller's palm are similar to those of the bird-of-paradise plant

- The vernacular name of the traveller's palm is said to be derived from the fact that the cup-like leaf bases hold water which travellers could drink

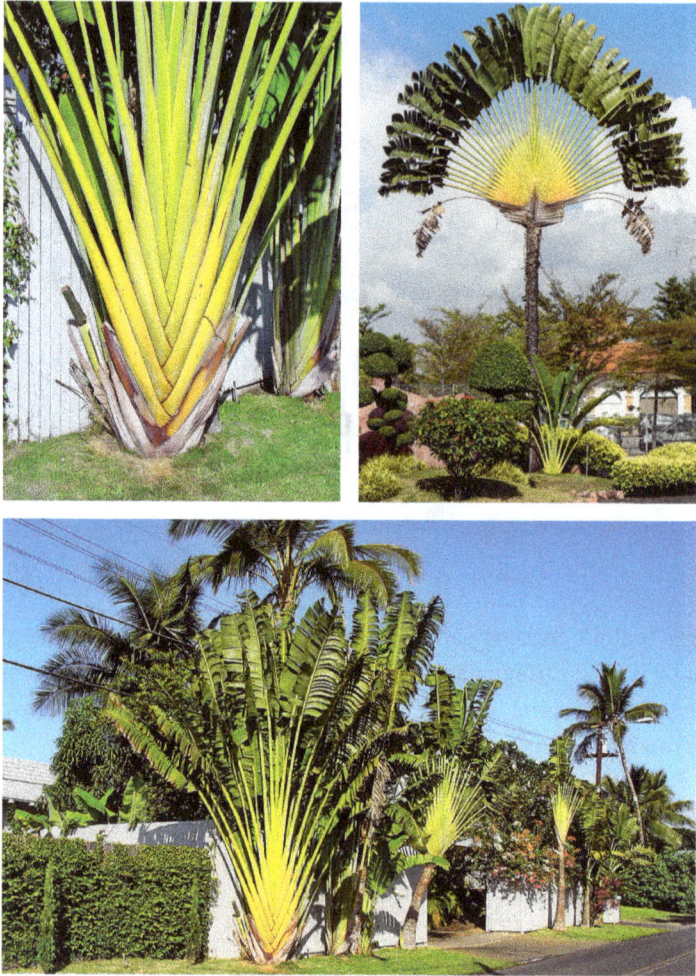

Sago palm (*Cycas revoluta*)

- A major confusion is associated with this common name because it refers to the true palm *Metroxylon sagu* as well as to the palm-like Asian cycad, in the family Cycadaceae

- Both the stem (which is sometimes branching) and the terminal crown of pinnate leaves of *Cycas revoluta* are similar to those of a true palm

- Leaves are stiff and borne as a rosette not singly as in palms; the male inflorescence resembles a cone, a key identifying character

- It is the most widely cultivated as cycad

- Edible starch, "sago," can be extracted from the stem of both *Metroxylon sagu* and *Cycas revoluta*

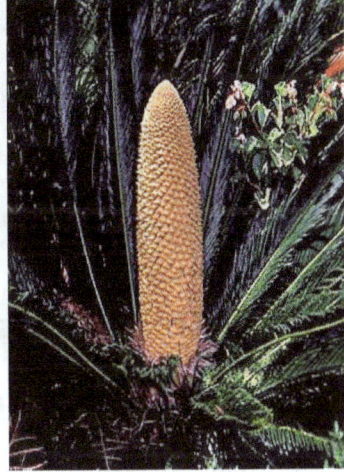

Palm lily or ti palm (*Cordyline australis* and *C. terminalis*)

- They are native to, respectively, New Zealand and East Asia

- They belong to agave family, Agavaceae

- The branching habit gives the 'palm lily' a resemblance to the branching palm *Hyphaene*, but has sword-like leaves crowded together at the end of the branches

- These two species of *Cordyline* resemble plants in the genus *Dracaena*, with which they are often confused

Screw palm (*Pandanus spiralis*)

- Native to Australia and tropical Asia, its morphology somewhat resembles the branching palm (*Hyphaene thebaica*)

- It belongs to the family Pandanaceae

- The screw palm's sword-like leaves form tufted crowns and the tree bears large pineapple-like fruits

- Where *Pandanus* spp. occur, leaves are widely used for weaving mats, baskets and so on

Palm fern (*Cyathea cunninghamii*)

- This plant is indeed a fern and not a palm

- It belongs to the tree-fern family, Cyatheaceae

- Native of New Zealand and Australia

- It has a single stem and pinnate leaves and somewhat resembling those of a true palm

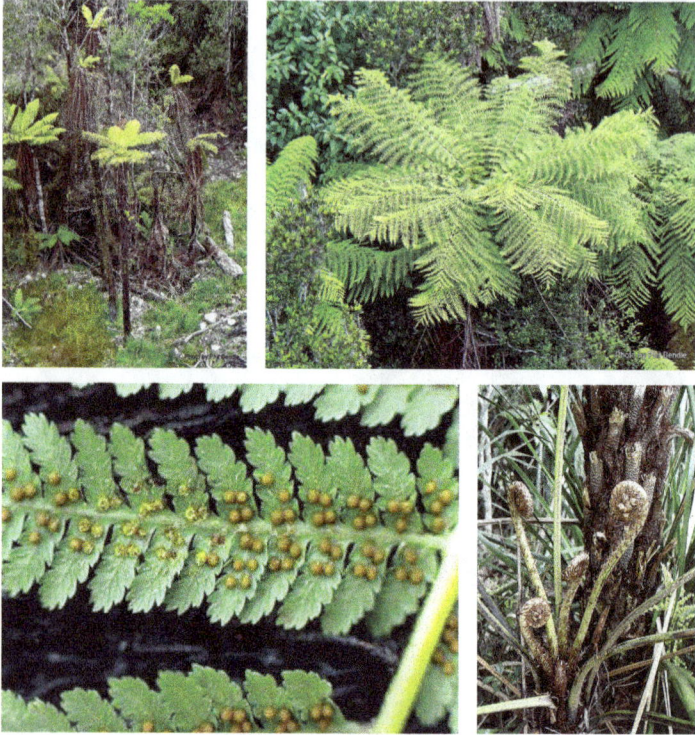

Palm grass (*Setaria palmifolia*)

- It is a perennial Asian grass

- It is an indication that the entire leaves resemble those of certain palms

- It belongs to the grass family, Gramineae/Poaceae

Panama hat palm (*Carludovica palmata*)

- It is a monocot plant and look like a palm

- It is a member of the Cyclanthaceae family

- It possess palmate leaf

- It is a stemless understory plant

- It is native of the lowland forests of central and south America, is often mistaken for a true palm

- The common name comes from the use of the fibre of young leaves to weave high quality hats

6

Description of Palms

1. *Acoelorrhaphe wrightii*

Common Names: Everglades palm, Paurotis palm, Saw cabbage palm, Silver saw palm, Silver saw palmetto

Description:

- It is a small to moderately tall palm that grows in clusters to 5 - 7 metres (16–23 ft), rarely 9 m (30 ft) tall, with slender stems less than 15 centimetres (5.9 in) diameter
- The leaves are palmate (fan-shaped), with segments joined to each other for about half of their length, and are 1 - 2 m (3.3–6.6 ft) wide, light-green above, and silver underneath
- The leaf petiole is 1 - 1.2 m (3.3–3.9 ft) long, and has orange, curved, sharp teeth along the edges
- The flowers are minute, inconspicuous and greenish, with 6 stamens. The trunk is covered with fibrous matting
- The fruit is pea-sized, starting orange and turning to black at maturity
- The fresh seeds germinate without difficulty and seedling growth is moderately fast
- It is of salt tolerant

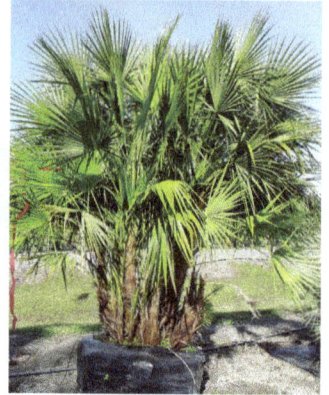

2.   *Adonidia merrillii*

Common names: Adonidia palm, Christmas palm, Dwarf royal palm, Manila palm

Description:

- It is an ornamental and evergreen palm tree
- It is a small to medium-sized, single-trunked palm tree growing moderately fast to a height of about 20 ft
- It's greyish stem is ringed by semi-circular leaf scars and topped by a 2-3 ft smooth and light green crown shaft that supports a crown of about a dozen pinnate (feathery) fronds
- The attractive arching fronds are about 5 ft long with deep-green strap-shaped leaflets that are about 2 ft long and 2 inches wide
- Fresh seeds germinate readily and further growth is also fast
- Prefers full sun though it can tolerate semi-shade

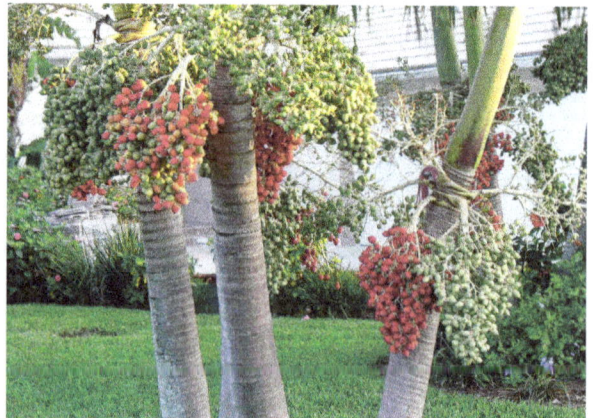

3. *Aiphanes aculeata* (syn: *A. caryotifolia*)

Common names: Chonta Ruro, Coyure Palm, Ruffle Palm, Spine Palm

Description:

- It is a solitary-stemmed, small spiny palm growing to a height of up to 10 m tall
- It has a very characteristic crown of spiny broad ruffled fronds
- Stem is pale grey, slim and armed with dark sharp spines
- Leaves are pinnately compound, armed with long sharp spines, leaflets are light green, broad at the tips
- Flowers are orange, yellow, subtended by a woody spathe, also armed with spines
- Fruit is one-seeded, round usually bright red, spherical and a single-seeded drupe
- The epicarp and mesocarp of the fruit is rich in carotene
- Seeds are used to make candles and are easy to germinate

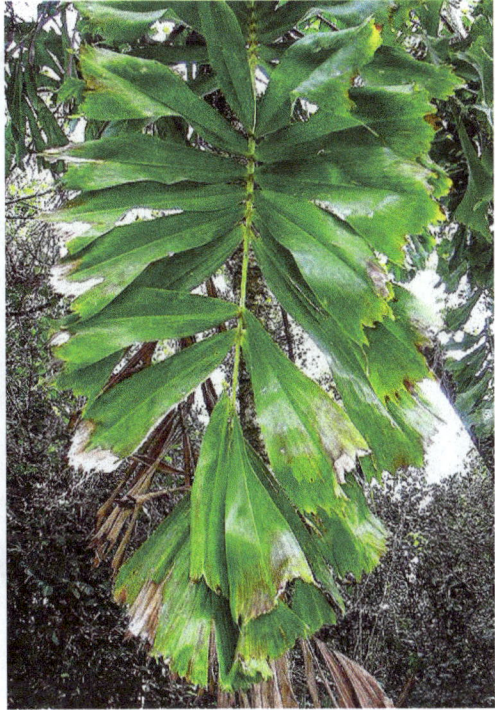

4.   *Aiphanes erosa*

Common names: Macaw palm

Description:

- It is a spiny palm growing to a height of about 5 - 8 metres tall
- Younger stems are covered with rings of black spines, but on older stems these are often lost
- Leaves are pinnately compound and leaflets emerge on either side of the axis of the leaf in a feather-like or fern-like pattern
- The lower surface of the leaf can be covered with spines up to 3 cm long or can be unarmed; the upper surface has a row of spines about 1 cm long along the midrib
- The rachis can be unarmed but is often covered with black spines up to 6 cm long
- It is widely planted as an ornamental plant
- The endosperm of the seed is edible, and is similar in taste to that of a coconut

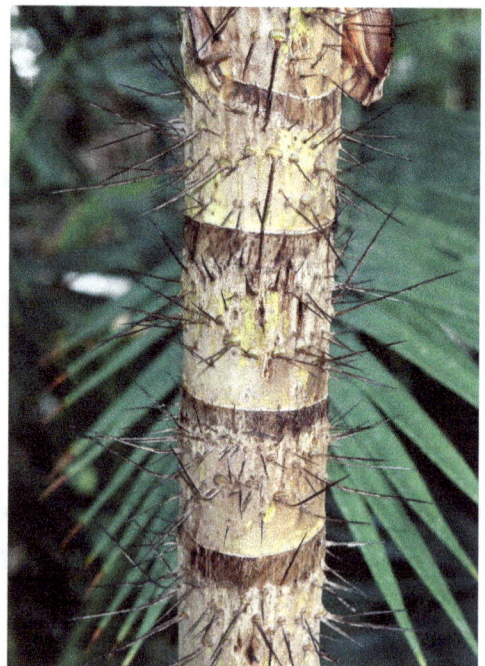

5. *Archontophoenix alexandrae*

Common names: Alex palm, Alexandra king palm, Alexander king palm, King palm, Northern bungalow palm

Description:

- Trunk is grey colour, up to about 30 m tall, fast growing palm
- Leaves are whitish or ash-colored on lower surface, sheath green
- Inflorescences are up to 70 cm long, with pendulous rachillae and are of white colour
- Fruits are pea-sized and of red colour
- Seeds are small and quick to germinate
- Used as a houseplant in humid location with bright light

6. *Archontophoenix cunninghamiana*

Common names: Bangalow palm, Piccabeen bungalow palm, Piccabean palm

Description:

- This palm looks similar to *Archontophoenix alexandrae* in both appearance and size and cold tolerant
- The leaves have paler stalks and are bright green above and below
- They are easily damaged by heavy winds
- They are often best suited to more sheltered areas
- Flowers are violet coloured

- Fruits are red in colour and are attractive to birds
- The small seeds germinate quickly

7. *Areca catechu*

Common names: Arecanut palm, Betel nut palm, Betel palm, Caccu, Catechu, Pinang

Description:

- It is a medium sized palm, growing to a height of about 20 m tall
- Trunk is solitary, slender and erect with 10–15 cm diameter
- Flower stalk is white, sweet scented
- Fruits are orange-yellow when ripe and about 2.5 inches long
- The leaves are 1.5–2 m long, pinnate, with numerous, crowded leaflets
- The seed contains alkaloids such as arecaidine and arecoline, which, when chewed, are intoxicating and slightly addictive
- Nut itself is brown, oval and flattened at one end
- It is popular for chewing and also used in interiorscaping
- It is often used in large indoor areas such as malls and hotels
- It is of slow growing, needs less water and more light

8. *Arenga caudata*

Common name: Dwarf sugar palm

Description:

- It is a low growing, suckering palm with dense, compact, clumping growth habit
- Stems are multistemmed which are growing up to 1.5 m tall
- Leaves are green to dark green on the upper surface, silverywhite on the lower surface, ending in a triangular (wedge-shaped) leaflets
- Inflorescence are solitary
- Fruit is oval to oblong looks like pea and is of red colour at maturity

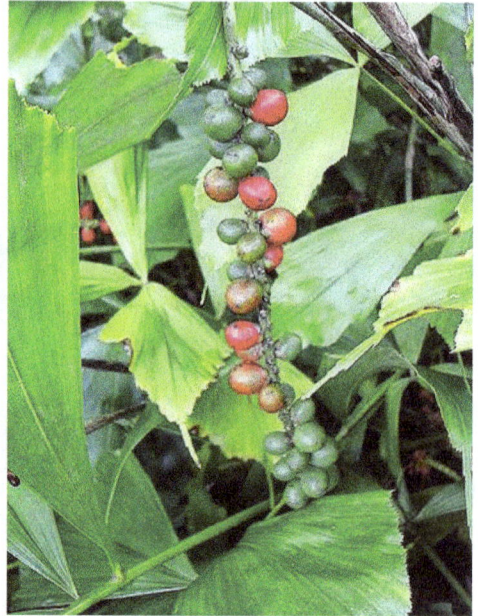

9.   *Arenga engleri*

Common name: Dwarf sugar palm, Sugar palm, Formosa palm, Taiwan Arenga

Description:

- It is a clustering palm that is multi stemmed
- It has dark green leafs with slender, clean, distinctly ringed trunks
- It produces suckers, also known as basal offshoots; these suckers grow to their maturity and then replace the older stems as they die
- The fruits are red and yellow and seeds germinate readily
- It doesn't tolerate drought conditions but withstand frost
- It is used for following purpose in landscaping
    - Privacy screen along a patio, pool or anywhere
    - Hedge plant
    - Single specimen
    - Accent for corners
    - Covering a fence
    - Container palm

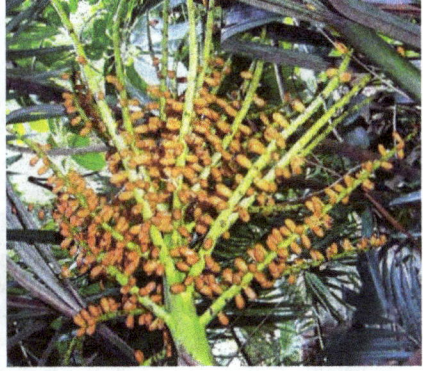

10. *Arenga pinnata*

Common name: Aren, Areng palm, Black fibre palm, Gomuti palm, Kabong, Sugar palm

Description:

- It is a solitary, unarmed, monoecious feather palm
- It is a medium-sized palm, growing to 20 m tall, with the trunk remaining covered by the rough old leaf bases
- The leaves are 6-12 m long and 1.5 m broad, pinnate and are used as thatch material
- The fruit is subglobose, 7 cm diameter, green colour turns to purplish black on maturity with 2-3 seeds

11. *Arenga undulatifolia*

Common name: Aren Gelora

Description:

- It is a monoecious plant, monocarpic (bearing fruit only one time during its existence)
- It is a fast growing clustering palm
- Leaves are pinnate with wavy edges
- Trunk is caespitose (growing in tufts or clumps) and hidden by old leaves
- Flower: Spathe emerges from crown
- Fruit: Dark red when ripe

12. *Bactris gasipaes*

Common names: Chonta, Peach palm, Pejibaya, Pejibeye, Pejivalle, Pewa, Pupunha

Description:

- It is a caespitose palm with an extensive but fairly superficial root system
- It is a multi-stemmed tree that grows up to 20 m tall
- The internodes of the stems (stem regions between leaves) and the leaf sheaths are armed with spines that can be up to 5 cm long
- The trunk has internodes covered with spines, alternating with nodes without spines, formed by leaf scars
- The inflorescence is covered with two bracts
- The fruit occurs in a drupe of variable size – 300 to 400 g

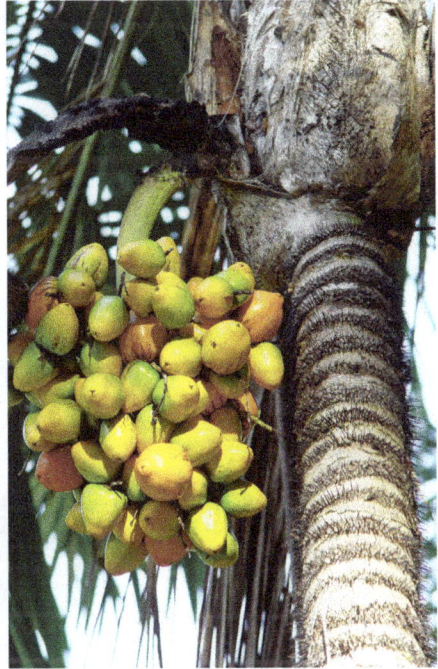

13. *Bismarckia nobilis*

Common name: Bismarck palm

Description:

- It grows from solitary trunk, grey to tan in color, which show ringed indentations from old leaf bases

- Trunk is 30 to 45 cm in diameter, slightly bulging at the base, and free of leaf bases in all but its youngest parts
- In their natural habitat they can reach above 20 meters in height but usually get no taller than 12 m in cultivation
- Mostly leaves of this type is silver-blue colour
- The oval fruits, which hang down from the leaves, are 1.5 to 2.0 inches long and brown when ripe
- The seeds are easy to germinate

14. *Borassus flabellifer*

Common names: Doub palm, Lontar plam, Palmyra palm, Pannaimaram, Tala palm, Talauriksha palm, Tal-gas, Toddy palm, Wine palm

Description:

- It is a robust tree and can reach a height of 30 metres
- The trunk is grey, robust and ringed with leaf scars, old leaves remain attached to the trunk for several years before falling cleanly
- The leaves are fan-shaped and 3 m long, with robust black teeth on the petiole margins
- It is dioecious palm with male and female flowers produce on separate plants
- The fruits are black to brown with sweet, fibrous pulp and each seed is enclosed within a woody endocarp
- It prefers hot, sunny, well drained location and drought tolerant
- The fruit is round, large, 6-8 inches in diameter and containing three seeds, is edible

- It is grown in gardens and parks as landscape palm species
- The cut flower stalks yield sap, which is turned into palm sugar or fermented into toddy, the timber is used for construction

15. *Brahea armata*

Common names: Blue fan palm, Blue hesper palm, Grey goddess, Mexican blue fan palm, Mexican blue palm, Short blue hesper

Description:

- It grows to a height of about 15 m or more
- It has a stout trunk
- It produces bluish leaves and are of 1-2 meters wide, with a meter-long petioles
- The leaves are persistent in nature, forming shag around the trunk
- The inflorescence extend out beyond the crown, reaching 5 meters in length
- Fresh seeds germinate easily

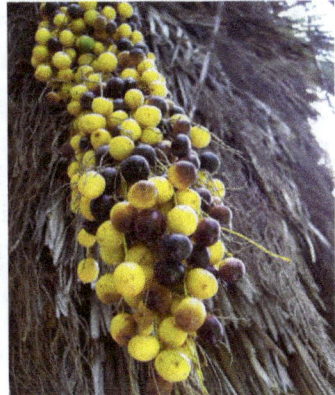

16. *Brahea edulis*

Common names: Guadaloupe palm, Guadalupe palm

Description:

- It is a fan palm which grows to 4.5–13 meters tall
- It grows between 400 and 1000 meters above mean sea level
- It is suitable for both temperate and dry subtropical climates
- It shows a good resistance to drought, cold, and frost
- The presence of the goats prevented regrowth of the native trees
- The round fruits are produced in huge quantities and germination is quite good

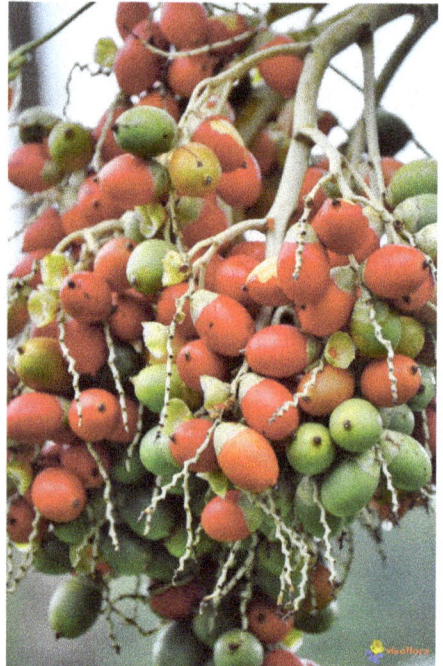

17. *Butia capitata*

Common names: Butia palm, Jelly palm, Pindo palm, South American jelly palm, Wine palm

Description:

- It is a cold-hardy, single-trunked palm

- It has a rounded canopy of blue-grey, strongly-recurved and graceful fronds which curve in towards the trunk
- The heavy, stocky trunks are covered with persistent leaf bases
- It produces large, showy clusters of orange-yellow, juicy, edible fruits and are produced and often used to make jams or jellies
- Popular as a garden palm, it can stand frost when mature
- Seeds germinate easily when heat, humidity are provided

18. *Calamus australis*

Common names: Hairy mary, Lawyer cane, Lawyer's cane, Rattan palm, Wait-a-while palm, Wait-a-while vine

Description:

- A slender vine not exceeding a stem diameter of 2 cm. Vine stem is smooth and glossy
- Leaves are compound and grows up to 1 m long
- Young inflorescence consists of female and sterile male flowers. The sterile male flowers are aborted so that the mature inflorescence contains only female flowers
- Fruits are globular
- Small seeds lose their viability very quickly
- Seeds to be sown fresh for good germination

19. *Carpentaria acuminata*

Common name: Carpentaria palm

Description:

- It is a single-stemmed palm growing up to 9-30 m tall on a single smooth grey trunk
- The attractive crown is composed of 10-12 gracefully arching leaves, which are deep green on top and blue-green below
- Leaves are pinnate and are of 2-4 m long
- Inflorescence arise from the upper leaf bases and then progressively down the trunk
- Both male and female flowers occur on the same inflorescence and are green to white in colour
- Fruit is globular and red in color
- Seeds to be sown fresh for quick germination

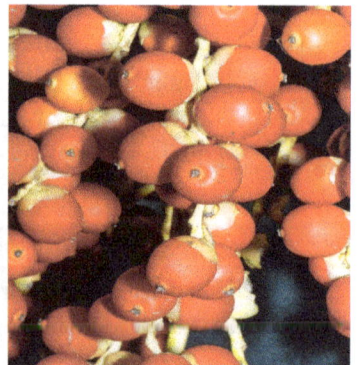

20. *Caryota mitis*

Common names: Burmese fishtail palm, Clustered fishtail palm, Fishtil palm and Tufted fishtail palm

Description:

- It forms multi-stemmed clumps grows up to 8 m high and 4 m wide
- Each slender stem is topped with several bipinnate leaves than can reach up to 3 m in length
- The light green leaflets are shaped like a fish's tail fin, hence its common name
- It also has a unique way of flowering, the first flowering mop-like cluster emerges from the top of a mature palm and subsequent clusters emerge below and so on
- When the cluster reaches the ground, the palm dies
- The fruits contain caustic crystals, to be handled with care
- The black seeds to be removed from the flesh carefully and sown fresh

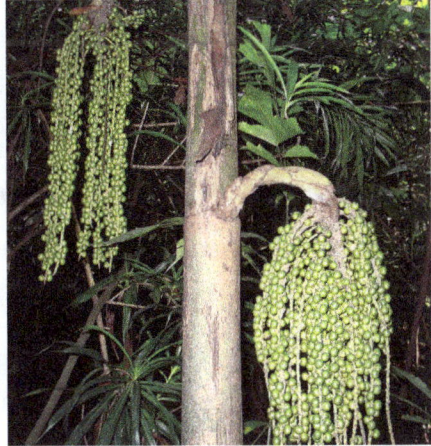

21. *Caryota urens*

Common names: Fishtail palm, Jaggery palm, Kitul tree, Sago palm, Solitary fishtail palm, Toddy palm, Wine palm

Description:

- It is a solitary-trunked palm that grows up to 12 m in height with 30 cm wide

- Widely-spaced leaf-scar rings covered its grey trunk which culminates in a 6 m wide and 6 m tall leaf crown

- The bipinnate leaves are triangular in shape

- The plants have a determinate growth habit; no new leaves originate after emergence of the 1st terminal inflorescence, which signals the start of the plant's reproductive phase

- Flowering begins at the top of the trunk and often continues downwards for several years

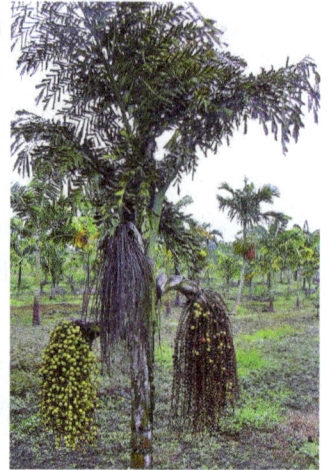

22. *Ceroxylon quinduiense*

Common names: Quindio wax palm

Description:

- It has an extremely slow growth and can live up to 100 years

- It can grow to a height of 50 to 60 m and it is the tallest palm in the world

- The trunk is cylindrical, smooth, light colored, and covered with white wax

- When the leaves die, they fall and this forms a dark ring around the trunk

- The fruits are orange-red and smooth when ripe

- It is of cold hardiness nature

23. *Chamaedorea cataractarum*

Common names: Cascade palm, Cat palm, Cataract palm

Description:

- It is a small, attractive, trunkless, clumping palm growing to a height of 2 to 2.5m
- Leaves are glossy with dark green leaves, and long thin leaflets
- As the trunk grows, it creeps across the ground helping to anchor the plant
- This trunkless habit, along with its flexible long thin leaflets is an adaptation which helps to prevent it from being washed away during floods
- It has become popular as a house plant

24. *Chamaedorea elegans*

Common names: Good luck palm, Parlor palm, Parlour palm

Description:

- It grows to 2–3 m tall with slender, cane-like stems
- It is excellent as a small palm for a terrarium or bonsai
- It is a dioecious palm which produce male and female flowers on separate plants
- The small seeds germinate easily, when sown fresh

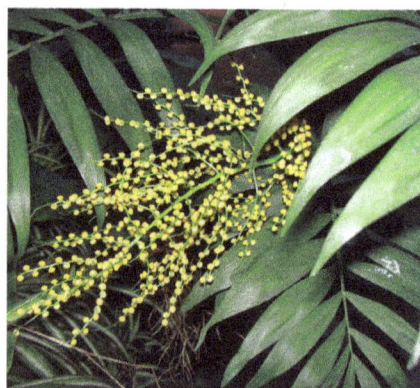

25. *Chamaedorea erumpens*

Common name: Bamboo palm

Description:

- It is a small, delicate, multiple-trunked palm

- It produces clumps of bamboo-like canes
- Leaves are pinnate and drooping
- New stems continually form at the base of the plant
- Since older leaves die and hang onto the stem, they require manual removal to keep the plant looking neat
- The individual dark green leaflets are almost papery, and the last few leaflets at the tip of the leaf are several times wider than others on the leaf
- Flower is white colour
- It needs rich, well-drained moist soil and a shady location
- It is suitable to grow in containers as a house plant
- It makes a delicate, fine-textured accent in a shrub border or in a low-growing groundcover

26. *Chamaedorea metallica*

Common names: Metallic palm, Miniature fishtail plam

Description:

- It is of slow growing nature
- It grows to a height of up to 5-10 ft tall and 1-5 ft wide
- It has a single thin green trunk ringed by old leaf scars
- Leaves are metallic blue-green, pinnately veined, and resemble a fishtail shape, hence the common name dwarf fishtail palm
- It produces small red, purple or orange flowers that grow in a branched inflorescence

- Fruits are green colour and egg-shaped which turn black when ripe
- Tolerate low light conditions extremely well

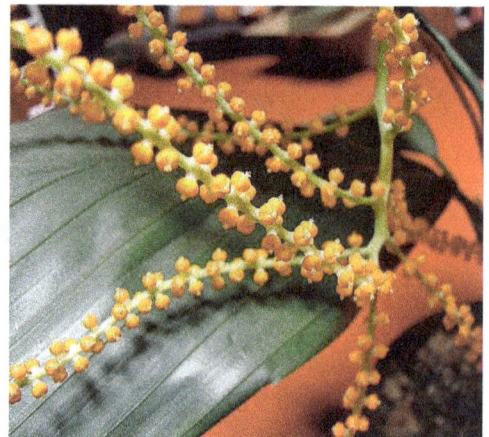

27. *Chamaedorea microspadix*

Common names: Bamboo palm, Hardy bamboo palm

Description:

- It is a hardy bamboo palm
- It is an elegant, clumping palm which grows to a height of about 3 m
- Leaves are pinnate with slightly drooping
- Fruits are spherical in shape and of orange-red colour when ripe
- The small round seeds germinate readily if freshly sown

28. *Chamaedorea seifrizii*

Common names: Bamboo palm, Reed palm

Description:

- It is a wonderful palm will grow indoors to a height of about 4-6 feet
- It forms clusters of stems each of which is very narrow
- It is recommended as air purifying plant for indoors by NASA
- Each stem is long and slender with "nodes"
- At the base of the stem, just above the soil surface small roots called adventitious roots can often be seen
- It naturally spreads by suckers or offshoots
- Leaves are dark green, pinnate and as the old fronds die, these should be trimmed off
- It prefers shade, can withstand higher light
- The flowers are dull yellow in color
- The fruits are usually small pea-sized berries that are orange/red in color

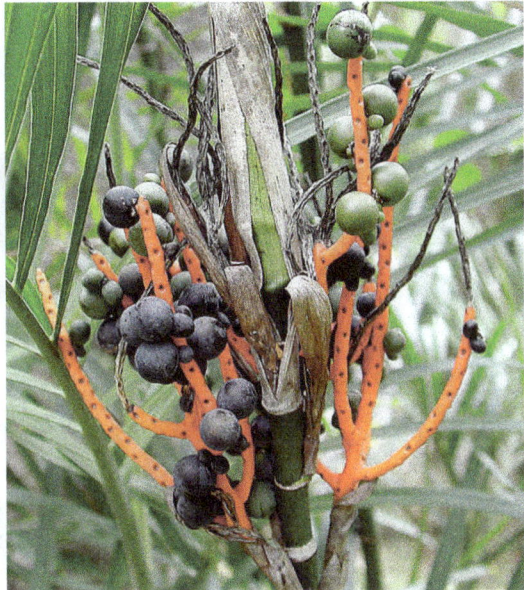

29. *Chamaedorea stolonifera*

Common name:　　Reed Like Palm

Description:

- Leaves are pinnate, 5 to 8 simple leaves, bifid to 2/3 of their length (divided at the tip)
- Trunk are clustering type
- No crownshaft, green with swollen leaf rings
- Flower stalk coming from the lowest leaves
- Fruits are black in colour, 0.3 inch in diameter (8 mm)
- Seeds are oval in shape

30. *Chamaerops humilis*

Common names:　　European fan palm, Fan palm, Mediterranean fan palm

Description:

- It is a small clustering palm, growing to a height of 1-4 metres, rarely to 6 metres
- Leaves are palmately compound, 1-1.5 m long, with 10-20 fingered leaflets of 50-80 cm long arranged in a fan at the end of the heavily armed 30-70 cm petiole

- *C. humilis* var. humilis, the leaves are green, while in *C. humilis* var. argentea they are strongly glaucous with a silvery-blue waxy coating
- It is a monotypic genus and dioecious flowering habit
- It also has numerous sharp needle-like spines produced on the leaf stems; these may protect the stem growing point from browsing animals
- The flowers are borne in dense, short inflorescences at the top of the stems
- Unripe fruits are bright green, turning to orange to brown when ripe
- The seed (usually 0.6–0.8 g) comprises a small cylindrical embryo, which is surrounded by several layers, from inner to outer
- It has an underground rhizome which produces shoots with palmate, sclerophyllous leaves
- Due to its rusticity and resprouting ability after fire, it has a high ecological value for preventing erosion and desertization
- The young unopened leaves are treated with sulphur to make them softer and are then used for finer work
- The leaves of the adult plants have been used in basket weaving to make mats, carrier baskets, and brooms
- The fruits are not edible but have been traditionally used in medicine as an astringent because of their bitterness and high tannin content
- The palm is very hardy to cold and suitable for temperate gardens

31. *Coccothrinax argentata*

Common names: Florida silver palm, Silver palm, Silver thatch palm, Silvertop, Silver top palm

Description:

- It is a medium sized fan palm growing about 10 m tall

- It is also commonly known as thatch palm because its leaves are used for weaving hats and baskets

- Trunk is solitary, or rarely clustered, grey to blackish, slender, up to 20 cm diameter, covered in a woven thatch of fibres smooth in the older basal parts

- Leaves are circular, palmate (fan-shaped), deeply lobed, dark green above and silvery coating below

- The inflorescences are short, branched and grow out just below the leafy canopy

- Flowers are tiny yellowish or ivory colour

- Fruit is round and is of purple-black in colour

- It tolerates sweltering heat and windy conditions and moderately salt tolerant

- It is always slow-growing, taking decades to reach tree-like heights

- It is grown as an ornamental plant in gardens in tropical and subtropical regions

- It is also used medicinally by traditional healers to treat uterine fibroids and hot flashes

32. *Coccothrinax crinita*

Common names: Mat palm, Old man palm, Old man thatch palm, Palma petate, Thatch palm

Description:

- It is a solitary, medium-sized, palmate palm
- It is a single-stemmed palm which grows to a height of about 2 to 10 meters tall with stems of 8 to 20 centimetres in diameter
- This plant produces flowers which are small; light yellow in colour and are not showy
- These tiny flowers usually cluster on a long stalk that droops down from the canopy; this display can be showy
- Fruits are fleshy, 0.7–2 cm in diameter and is of black and purple in colour
- It prefers partial or full sunlight
- It can tolerate drought
- Fibres of this palm are used for pillows, the trunk for shelter, and the leaves for bowls

32. *Coccothrinax spissa*

Common names: Guano, Swollen silver thatch

Description:

- It is solitary, fan-leafed palm with bulbous white trunk growing to a height of about 3 to 8 metres tall
- Leaves are silvery underneath
- The fruit is dark purple and is of 1.1 to 1.2 cm in diameter

- It grows well in slightly acidic to alkaline soils
- Its growth rate is slow

34. *Cocos nucifera*

Common names: Coco palm, Coconut palm, Coconut

Description:

- It is a long-lived palm that may live as long as 100 years
- It has a single trunk, very strong and elastic growing to 20-30 m tall
- Bark is smooth and grey, marked by ringed scars left by fallen leaf bases
- The leaves are 4 to 6 m long, pinnate
- It is used for making thatching, hats, baskets, furniture, mats, cordage, clothing, charcoal, brooms, fans, ornaments, musical instruments, shampoo, containers, implements and more
- The most sterile water on earth is found in this nut

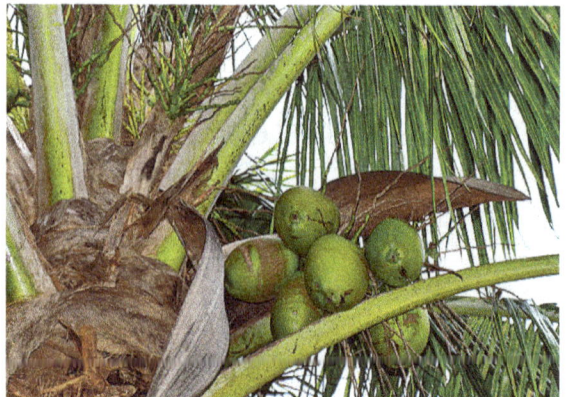

35. *Copernicia baileyana*

Common names: Bailey fan palm, Bailey's copernicia palm, Yarey, Yarey Hembra, Yareyon

Description:

- It is a solitary, large, fan-leaved palm growing to a height of about 10-20 metres tall
- Trunk is swollen, white colour with 60 cm in diameter
- Leaves are bright green and deeply segmented
- Fruit is black, 1.8 to 2 cm in diameter
- It prefers alkaline to neutral soils
- Its growth rate is slow to moderate
- Leaves are used for weaving hats, baskets and also for thatch making

36. *Copernicia macroglossa*

Common names: Cuban petticoat palm, Jata de Guanbacoa, Petticoat palm

Description:

- It produces solitary trunk which grows up to 9.1 m high, and 20.3 cm in diameter
- Leaves are upright fan-shaped with no petiole that grow in a spiral formation along the top of the trunk
- From the bottom to upper stem, a beard like structure made out of dry fan shaped leaves extend to cover approximately half of the trunk which gives the appearance of petticoat, hence its named

- It is a dioecious plant which develops their female and male flowers on different plants
- The flowers are white followed by 1 inch black fruits that are usually oval in shape

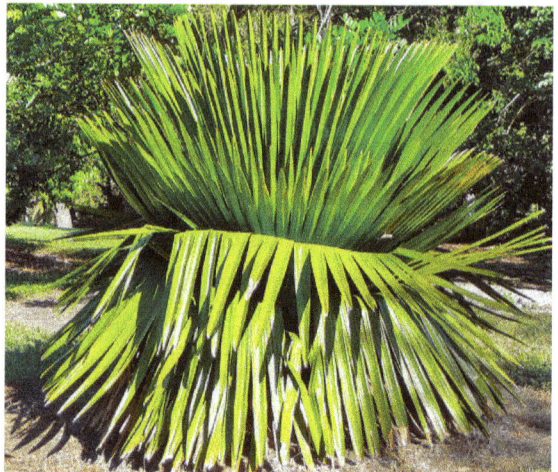

37. *Corypha umbraculifera*

Common names: Talipot palm

Description:

- It is one of the largest palm growing to a height of about 25 m
- It is a fan palm with large, palmate leaves of up to 5 m diameter which can shelter 10 men from rain
- It bears the largest inflorescence of 6-8 m long, consisting of one to several million small flowers borne on a branched stalk that forms at the top of the trunk
- It is a monocarpic palm that flowering only once in its life time of 30 to 80 years old
- It produces thousands of round, yellow-green fruit of 3-4 cm diameter
- Fruit can take about a year to mature and each containing a single seed
- The plant dies after fruiting
- The leaves are also used for thatching and the sap is tapped to make palm wine
- A large garden is ideal for growing this palm

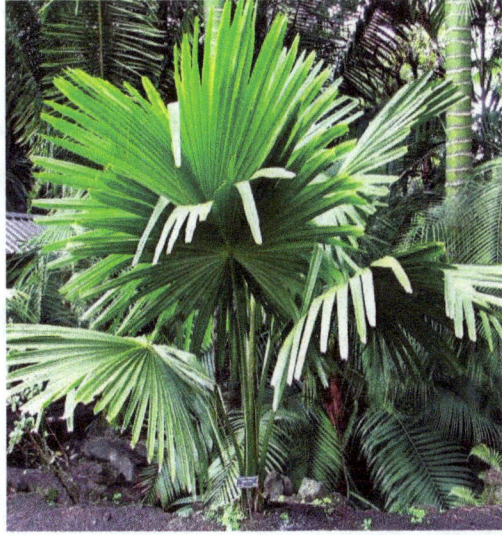

38. *Cyrtostachys renda*

Common names: Sealing wax palm, Lipstick palm, Maharajah palm

Description:

- It is a slender, clustering palm with multiple stems arise from the base growing to a height of about 15-20 m height

- Stems are 5-14 cm in diameter, green with greyish stripes or yellow with somewhat greenish and purplish stripes, internodes are 15 - 24 cm long, crown appearing as a shuttle-cock shaped

- The leaves are compound, each leaf contains around 50 leaflets, each of which is long and narrow can be up to 45 cm long

- The leaf bases, forming the crownshaft, the stems, and the leaf stalks (petioles) are a bright red color

- It is monoecious palm

- The flowers are found in clusters that are 60 cm long and that are visible beneath the green foliage

- The fruits are oblong in shape, up to 1 cm long, and are black with a scarlet base when mature

- Stems are used for making flooring and leaves for thatch

39. *Dictyosperma album*

Common names: Common princess palm, Hurricane palm, Princess palm

Description:

- It grows to a height of 30 feet on a single grey-ringed trunk
- The leaves are pinnate, 8-12 feet long and the leaflets are 3 feet long
- The erect young spear leaves look like long swords
- When young, the top of the fronds twist and droop because of the flexible midrib which gives a curving appearance to the leaves
- The crown shaft is also distinctive as it is light green, white or red and covered in grey to brown scales
- The inflorescence appears below the crown shaft in a large branching cluster
- The flowers are yellow to red
- The fruits are bullet shaped and purple or black in color
- It is obviously a useful palm for hurricane-prone areas near the coast
- It also makes a very graceful looking specimen or accent tree and even may be used indoors when young

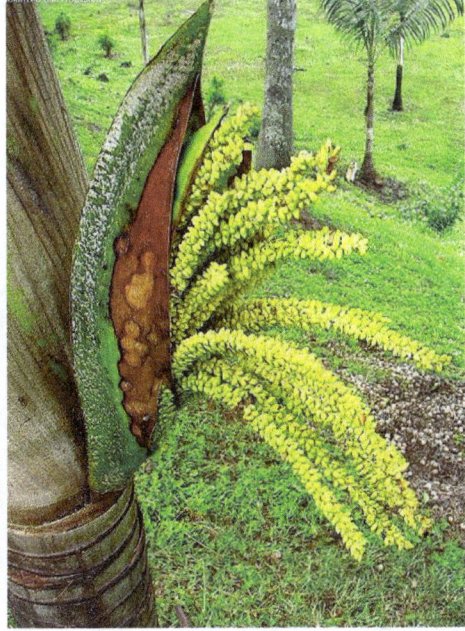

40. *Dypsis decaryi*

Common names: Three-sided palm, Triangle palm

Description:

- It is a solitary, monoecious palm with smooth, upright trunk, 22-32 cm in diameter, brownish-grey, ringed by the scars from the fallen fronds

- Leaves are pinnate or feather like, arching almost upright, about 3 m long and 1 m wide, segmented, bluish-green above and beneath supported by brown petiole covered in a whitish bloom

- Overlapping leaf-bases grow from three distinct points of the trunk about 120 degrees apart on the main stem, forming a triangular shape in cross section, hence it is named as triangle palm

- The inflorescences branch out from the axils of the lower leaves. They produce yellow and green flowers

- Fruit is round and black colour with 1inch in diameter

- It is a fine ornamental plant and is most valuable as a specimen tree or accent plant

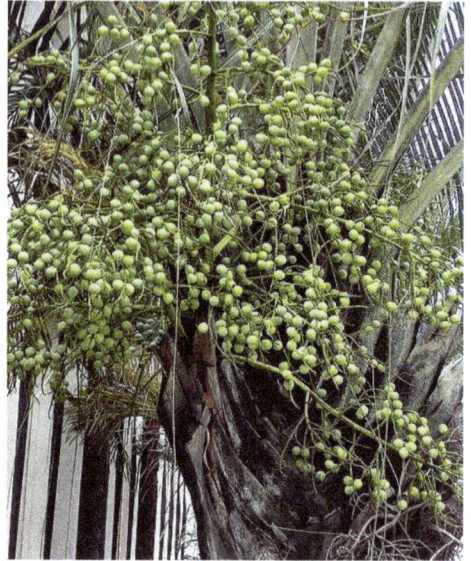

41. *Dypsis lutescens* (Syn: *Chrysalidocarpus lutescens; Areca lutescens*)

Common names: Areca palm, Butterfly palm, Cane palm, Golden feather palm, Golden cane palm, Madagascar palm, Yellow palm, Yellow butterfly palm

Description:

- It grows to a height of 6-12 m tall
- It produces multiple stems from the base
- The leaves are arched, 2-3 m long and pinnate, with 40-60 pairs of leaflets
- Flowers are yellow in colour borne on panicle inflorescence
- It is grown as an ornamental plant in gardens in both tropical and subtropical regions, and elsewhere indoors as a houseplant
- It purifies the air by filtering xylene and toluene and act as a humidifier
- The bunches of fruits, when ripe are yellow, produced profusely

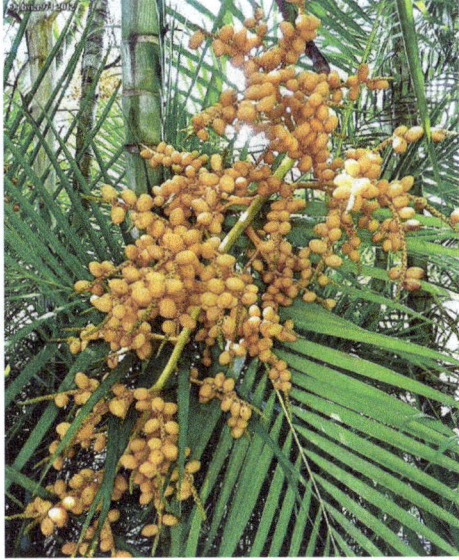

42. *Elaeis guineensis*

Common names: African oil palm

Description:

- It is a single-stemmed palm which grows to a height of about 20 m tall, important as a commercial crop
- The leaves are pinnate, spirally arranged, paripinnate, spinescent, clasping the stem at base
- The flowers are produced in dense clusters; each individual flower is small, with three sepals and three petals
- Fruit is reddish, about the size of a large plum, and grows in large bunches
- Plant extracts is also reported to be a folk remedy for cancer, headaches, and rheumatism
- Leaves are used for thatching; petioles and rachis for fencing
- It is also occasionally grown as an ornamental tree
- Kernel oil is used in the manufacture of soaps and detergents
- Press cake, after extraction of oil from the kernels, is used as livestock feed, containing 5-8% oil

43. *Euterpe edulis*

Common names: Assai palm, Jucara palm

Description:

- Stems are solitary, or rarely caespitose (growing in dense tufts or clumps)
- Number of leaves are 8-15 in the crown, spreading; olive green to dark green, sometimes reddish or orange-tinged, glabrous or with reddish brown scales
- Fruits are globose, 1-1.4 cm in diameter
- Seeds are globose shaped
- Inflorescence appear below crown shaft
- Young plants are grouped in a pot and used as house plants

44. *Hedyscepe canterburyana*

Common names: Big mountain palm, Umbrella palm

Description:

- It is a very slow-growing palm and of monotypic genus, growing up to 10 metres tall in sub-tropical climate
- It has a slender, close-ringed trunk and a prominent silvery crown shaft
- Fruit are deep red when ripe, egg shaped and about 4 cm long
- Fruits can take up to four years to ripen
- Each fruit contains a single seed
- It also does well in containers or as indoor plants where light is good

45. *Howea belmoreana*

Common names: Belmore sentry palm, Curly palm, Sentry palm

Description:

- It is a monoecious, pinnate palm and has a closely ringed green trunk, slim and solitary
- Leaves are curved with erect leaflets nearly forming a half circle in their arching up, out and down
- The leaves are quite long, but the width of the entire palm is much less than the doubled length of the leaves because of their extreme recurving
- It grows more slowly and does not do well in a pot
- It is less tolerant to full sun, cold, wind and indoor conditions

- It tends to be more popular as an outdoor garden plant
- Fruits oval, 4 cm in length and red-brown when ripe and take long time to ripen

46. *Howea forsteriana*

Common names: Forster sentry palm, Kentia palm, Paradise palm, Sentry palm, Thatch leaf palm

Description:

- It has a canopy of about three dozen gracefully drooping leaves which produce an airy and poised look
- The trunk is swollen at the base and has slightly raised annular trunk rings
- The leaves are pinnate (featherlike) and grow up to 3.7 m long with thornless 1.2-1.5 m petioles
- The leaflets are like fingers, 0.8 m long and 5 cm wide; they bend downward in a graceful fashion
- Leaflets are dark green on top and lighter green on the bottom
- It produces an inflorescence of about 1.1 m long which consists of white flowers on 3-7 spikes which are fused at their bases
- Male and female flowers are produced in the same inflorescence
- Mature fruits are dull red and egg shaped, about 3.8 cm long
- Seeds are erratic in germination

- It tolerates and adapt to a wide variety of soils including those that are neutral, acidic, clayey and slightly alkaline

- They perform best in rich loamy soil with excellent drainage

- It is of traditionally slow growers, however regular fertilization promotes maximum growth

- It can withstand low light and drought

- Popularly grown as a house plant

47. *Hydriastele wendlandiana*

Common names: Florence falls palm, Latrum palm

Description:

- The trunk may be solitary or multistemmed growing to a height of more than 6 m tall

- The leaves are pinnate and widely varied. Apical leaflets fused together and is much broader than most of the lateral leaflets

- Flowers borne in an inflorescence which resembles a cat-o'-nine-tails and emerges from sheaths 20-25 cm long on the stem below the leaves

- Fruits are globular, calyx persistent at the base

- Seeds are globular

© J.L. Dowe

48. *Hyophorbe lagenicaulis*

Common names: Bottle palm

Description:

- It is popular due to its strange, swollen, bottle-shaped trunk
- It has a large swollen, smooth, grey trunk and is bulbous at the bottom, resembling a bottle when young
- As the tree matures, the width of the trunk evens out, appearing slightly conical in the oldest individuals
- It has only four to six leaves which will open at any time
- Each frond (leaves) arches upwards and curves down splendidly
- The flowers of the palm arise from under the crownshaft
- Flowers are white and numerous which grow on stalks at the crown of the tree
- Fruits are round and small
- Useful as houseplants or conservatory plants when small

49. *Hyophorbe verschaffeltii*

Common names: Spindle palm

Description:

- It has a spindle-shaped, swollen trunk growing to a height of about 6 meters
- Leaves are pinnate and stiffly ascending and lightly recurved
- Flower spikes are horn-like and emerge from below the crown shaft and are orange in colour and scented
- Fruits are oval, 3 cm long and bright red when ripe
- Seeds are beaked and spindle shaped
- It is of fast growing nature and making it ideal for the tropical garden or as an attractive indoor plant

50. *Hyphaene thebaica*

Common names: Doum palm, Branching palm, Gingerbread palm

Description:

- Branching habit is the unique characteristic of this palm
- The leaves are strongly recurved, blue green, costapalmate with thorny leaf stalks
- Fruits are 10 cm long, pear shaped and become orange when ripe and smells like gingerbread
- Can be planted as an accent or specimen tree
- Seeds need several days soaking before sowing
- Germination is erratic
- Very slow growing palm

51. *Johannesteijsmannia altifrons*

Common names: Diamond Joey

Description:

- It is a beautiful,s imple-leaved palm
- It is a medium sized, trunkless palm which grows up to 6 m in height
- Leaves are large, leathery, diamond-shaped with serrated edges and strongly pleated
- Inflorescence are erect initially and eventually pendent and later it get branched

- Flowers are cream coloured and pointed in bud stage
- Fruit is 4-5 cm diameter with long corky warts, brown when ripe
- The round seeds requires moist peat for sprouting
- Makes a good house plant and also hardy to cool conditions

52. *Jubaea chilensis*

Common names: Chinese wine palm, Honey palm, Honey wine palm, Little cokernut, Syrup palm, Wine palm

Description:

- It is the hardiest palm which grows to a height of about 15 - 25 m tall
- It possess thick, columnar trunk of 1.3 m wide
- Leaves are feather shaped and are dull green above and greyish underneath
- It is highly drought tolerant, cold- hardy but slow growing nature
- It takes one to two decades to get a large specimen plant
- It really deserves adequate space in the garden to thrive and look their best
- The sap of this species has been used to make wine and also soft drinks
- The actual seed (nut) is also edible and can be a source for palm oil
- The seeds are known locally as 'coquitos' or miniature coconuts, round, 2 cm in diameter, and are edible

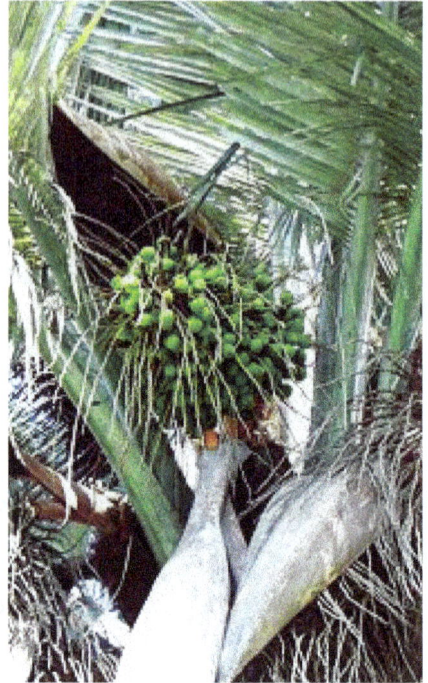

53. *Jubaeopsis caffra*

Common names: Pondoland palm

Description:

- It is a suckering coconut look-alike palm, grows to about 20 feet tall
- It is suitable for sun and also partial shade
- It is very rare in cultivation
- Seeds difficult to germinate

54. *Laccospadix australasica*

Common names: Atherton palm, Queensland kentia

Description:

- It is a solitary or clumping palm which grows to a height of about 7 m tall
- The trunks are clean and woody and are light tan color, 10 cm diameter in solitary specimen, but clumping plants are lower with smaller trunks
- The crown is brown and fibrous
- The leaves are pinnate (feather shaped). In case of solitary form, the emerging leaves are red in colour
- It has long, pendulous inflorescences and produce bright red fruits
- The fruits are small red and oval, growing along the length of the spike after flowering
- It makes an ornamental garden palm in a shady position
- Seeds lose their viability very quickly
- Suitable as houseplants and tolerate low light conditions

55. *Latania loddigesii*

Common names: Blue latan, Blue latan palm

Description:

- It is a solitary, erect palm growing to a height of about 10 m tall with undulating leaf scars
- Trunk is bulging at the base and measures up to 25 cm in diameter
- Leaves are pale waxy blue to blue-green with dense, pale, wooly tomentum on the underside and of 3 m width
- The leaf sheath splits to form a V-shape below the petiole
- Male and female flowers are produced on sepearte palms
- Inflorescence with male flower is shorter (up to 1 m long) and with female flower is long (up to 2 m long)
- Fruits look like peach and apricot fruit, dark brown when ripe
- Seeds sprout readily in a few weeks
- Suitable for tropical garden

56. *Latania lontaroides*

Common names: Red latan, Red latan palm

Description:

- It is a slow growing palm and reaches a height of 12 m
- Trunk is grey, lightly ringed, swollen base
- Leaves are costapalmate, 8 to 24, attractive as young plants
- Young leaves have a reddish coloration, armed leaf stems with wool and wax at the bottom

- Flower stalk is 0.9 to 1.8 m long
- Fruit is brownish green, 5 cm long and plum shaped
- Seeds are round at one end, pointed at the other (almond shaped)

57. *Latania verschaffeltii*

Common names: Yellow, Yellow latan palm

Description:

- It is a sun loving palm, grows to a height of 10 - 12 m
- It comes up well in warm temperate to suptropical areas
- Plants are intense yellow coloured at early age
- It needs excellent drainage
- Each fruit contains one to three seeds, plum-like

58. *Licuala grandis*

Common names: Fan-leaved palm, Palas payung, Ruffled fan palm, Vanuatu fan palm

Description:

- It is a solitary palm with upright stem growing to a height of 3 m tall and 5-8 cm in diameter
- It is ridged with leaf scars and some remnant fibers from leaf sheaths
- Leaves are costapalmate, with a stiff, undulating blade and forming a semicircle
- Leaf surfaces are deep green with bifid leaf tips
- Inflorescence is of 2 m long and branched
- Flowers are small, white and bisexual
- Fruits are small (1-1.5 cm), spherical, and bright red when ripe
- It is an attractive house plant, easily grown from seed

59. *Licuala orbicularis*

Common name: Parasol palm

Description:

- It is a shrub palm that grows to a height of 1 -2 m high
- It does not form a trunk and leaves grow directly from the underground root
- The attractive leaves are large, circular, divided or undivided, pleated

- Inflorescences initially erect, then curved with whitish hermaphrodite flowers spirally arranged

- The fruits are globose, of 1- 2 cm in diameter, cherry red when ripe

- It requires a semi-shaded to shaded position, sheltered from the winds, young plants grow very slowly

- The leaves are locally used for wrapping foods, manufacturing hats, umbrellas and other handicrafts

60. *Licuala spinosa*

Common names: Mangrove fan palm, Spiny licuala

Description:

- It is a very vigorous and attractive palm forming large clumps to about 3-4 metres high

- The leaves are split deeply to base into about 10 or 12 segments, with square and notched tips

- Leaf stalks thorny and should be handled with care

- Fruits are small, round and each contain one seed

- It prefers full sun, a lot of water and is more cold hardy than most *Licuala* species

- It also grows well in poorly drained soils

- It is grown as a houseplants

61. *Linospadix monostachya*

Common name: Walking stick palm

Description:

- It is a small, solitary, under-story palm and be protected from wind
- Wind turns leaftips brown, and leaves lose their lusture
- Leaves are unevenly pinnate, dark green to about 3m long
- Larger leaf with good appearance will be seen in shady areas
- Inflorescence is long
- Fruit looks attractive with red colour and of 8 mm diameter

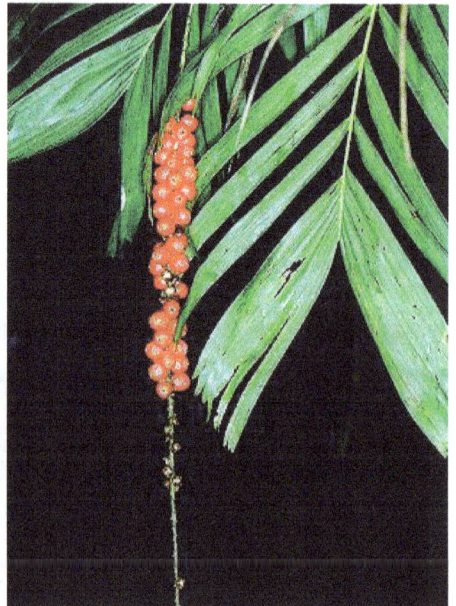

62. *Livistona australis*

Common names: Australian cabbage palm, Australian palm, Australian fan palm, Cabbage palm, Fan palm and Gippsland palm

Description:

- It is a moderately tall, fast growing fan palm popular in all places
- It is well suited to cultivation in tropical or warm subtropical areas
- It likes abundant water and grow well in waterlogged soils and areas with brackish water and coastal exposure
- While young, it makes a very pretty house plant, though the petiole contains thorns
- Stems are solitary, erect, grey or brown, to 15 m in height and 30-45 cm in diameter, but with age, it becomes bare with conspicuous, rough, closely spaced rings of leaf scars and vertical fissures
- Leaves are costapalmate, with a prominent hastula, in duplicate, glossy deep green, up to 2.5 m across
- Flowers are creamish yellow and have both male and female structures
- The fruit is spherical, about 2 cm across, and reddish-brown to black when mature
- Suitable for warm temperate to tropical climates

63. *Livistona chinensis*

Common names: Chinese fan palm, Chinese fountain palm, Fan palm, Footstool palm, Fountain palm

Description:

- It is a single stemmed, evergreen upright palm, grows up to 6 m tall

- Its mid green leaves are palm shaped, deeply divided and up to 1.5m across
- The leaf stalks are thorny
- Trunk has a diameter of up to 40cm
- Flowers are white and borne on 1.8m long inflorescence
- Fruit is blue/ black, fleshy and round, 1-2 cm in diameter
- It is grown as an attractive palm tree and a street tree
- Grown up palm is drought tolerant, suitable for tropical climates

64. *Lodoicea maldivica*

Common names: Coco-de-Mer, Double coconut, Seychellus nut

Description:

- It is a robust dioecious palm grows up to 25–30 m tall but grows very slowly
- Trunk of mature palm is simple, slender, with rings of leaf scars
- The leaves form dense crown at the top of trunk; petiole is 4–6 m long, blade is palmate, splitting into segments at the margin
- The male inflorescence resembles catkin, is terete, 1–1.5 m long and 10–15 cm in diameter
- Numerous scented flowers are developed in deep cavities
- Male and female flowers produced in separate plants
- The female inflorescence is 1–2 m long, individual flowers have 3 leathery tepals
- The fruit is a big ovoid drupe, usually with one stone up to 20 kg, covered by thin husk and stay on the plant for several years

- Seeds when sown, took several months to sprout after rooting, subsequent growth is slow
- Seeds are used as an aphrodisiac

65. *Lytocaryum weddellianum*

Common names: Dwarf coconut palm, Sago palm, Weddel palm

Description:

- It has a slim trunk, grow up to 5 m in height, slow growing in nature
- Leaves are finely pinnate with arching nature
- It is suited to a wide range of climates from cool temperate to cool tropical and can tolerate light frosts
- It grows best in a humus rich, acidic soil
- Seeds will show 100% germination, like coconuts, have three pores at one end

66. *Metroxylon sagu*

Common name: Sago palm

Description:

- It is a palm tree without leaf sheaths
- Leaves or fronds are pinnate, each month a new frond appears out of the growing point, and the oldest dies
- Dextrose sugar extract from sago palm starch can be processed to yield power ethanol
- The cortex of the trunk is also used for firing in paper mills. The bark may be used as a domestic fuel after drying
- Fibre resulted after processing the pith and also from leaves may be used for mat preparation
- The growing point and the young leaves around it may be used as a vegetable
- The inflorescence may contain up to 8,50, 000 fruits
- The seeds are large and germinates only under wet conditions

67. *Normanbya normanbyi*

Common name: Black palm

Description:

- It is a monoecious, arborescent, tall, handsome palm, growing about 20 metres tall, with four metres leaf spread

- The trunk is smooth, slender, and closely ringed, and becomes almost black as the palm gets older
- There is a pale green crownshaft, topped with a small head of feather like leaves
- It consists of many narrow leaflets, arranged circularly around the leaf stalk, which gives it a very bushy appearance
- It grows well in swampy areas in lowlands, up to 300 m elevation
- It has a silverish tinge on the underside of the leaves
- The green inflorescence comes from below the crownshaft
- Fruits are long, pear-shaped (5 cm), light-brown in colour

68. *Oncosperma tigillarium*

Common name: Nibung palm

Description:

- A slender palm that forms large and crowded clusters of tall, spiny trunks that can reach about 25 m (83 ft.) tall
- Trunk is topped with crowns of elegantly arching but spiny leaves with drooping leaflets
- Leaves are pinnate, 3 m long, leaflet many, drooping, 3 cm wide
- Inflorescence is panicle, 60 cm long, copiously branched
- Male flower is yellow, petals are lanceolate, apex cuspidate, 7 mm long
- Inflorescence is a panicle seen with 15–20 cm long peduncle, copiously branched

- Fruit is drupe, globose shaped with 1–1.2 cm diameter and turns black at maturity
- It is used for building construction, pig spears, palm heart is edible

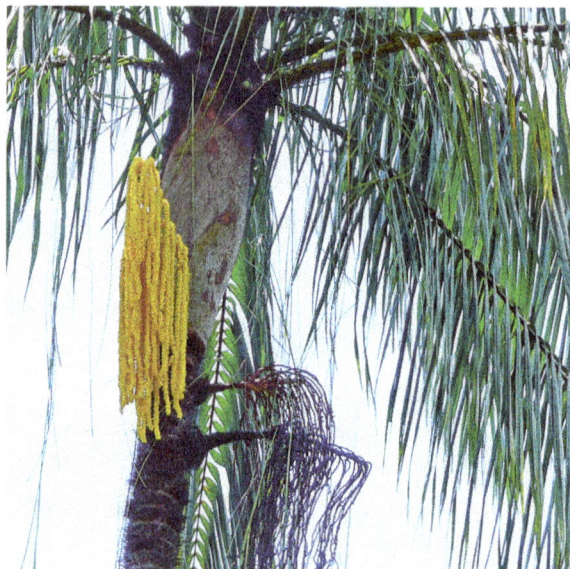

69. *Parajubaea cocoides*

Common name: Coco Cumbe, Coquito, Mountain coconut, Quito palm

Description:

- It is very hardy and wind tolerant
- It is a perfect palm for coastal climate

70. *Phoenicophorium borsigianum*

Common name: Latanier palm

Description:

- It is a very attractive solitary palm which grows up to 15m high, with a spined trunk
- Leaves are entire with very deep indentations and possess orange petioles
- It grows well in warm, sunny, moist and but well drained position
- The large dried leaves are used for thatching

71. *Phoenix canariensis*

Common name: Canary island date, Canary island date palm, Canary date palm

Description:

- It is an evergreen, tree-like palm
- The thick, hulking trunk is covered with interesting diamond designs that mark the point of attachment of the leaves
- The massive trunk supports a huge crown of over 50 huge arching pinnate leaves that may reach 18' long.
- Leaves are deep green shading to a yellow stem where the leaflets are replaced by spines
- It grows well in sunny position
- It does not comes up well in area with heavy soils and waterlogged condition
- Each large cluster of fruits (bright orange) appear on fruit stalks supported by a single leaf

72. *Phoenix dactylifera*

Common names: Date, Date palm

Description:

- It is a solitary, or sparsely clustering palm, with several suckering off shoots at base
- The trunk is 40 - 60 cm diameter, often with a much wider base
- The broad grey trunk is patterned with diamond-shaped leaf scars and grows to a height of 45.7 m tall
- Leaves are large greenish or bluish grey pinnate, are typically 5.5-6.1 m long
- It has long, sharp spines at the bases of the leaves, which are formed from modified leaflets
- Fruit is yellowish orange to red colour, called 'dates', oblong and about 3.8 cm in length
- It is also grown as an ornamental tree
- Male and female flowers are produced on separate palms, one male tree is planted for every 20 females

73. *Phoenix reclinata*

Common names: African date palm, African wild date palm, Senegal date palm

Description:

- It is a large, clumping, dioecious date palm that can grow to over thirty feet tall
- The leaves are long, pinnately divided and have long, sharp spines at the base of the leaf stalk
- Old leaves remain on the plant unless they are trimmed off
- Clusters of cream-colored flowers are produced among the leaves
- Fruits are small, orange to brown colour

74. *Phoenix roebelenii*

Common names: Dwarf date palm, Miniature date palm, Pygmy date palm, Roebelin palm

Description:

- It is a small, dioecious palm that can grow to ten feet tall or more
- It possesses solitary trunk but clumping types exist, too
- The flowers are small, creamy yellow
- The fruits are dull red-brown when ripe
- It grows well in a sunny site
- It seems very adaptable to different soil conditions
- Young plants are grown as house plants

75. *Phoenix rupicola*

Common names: Cliff date palm, East India wine palm, India date palm, Wild date palm

Description:

- It is a cliff date palm, growing to only about 20 ft tall
- It is smaller than many other common palms
- As the leaves die they shed leaving a smooth trunk ringed with narrow scars at the point of attachment
- The leaves are long (3.1 m) with bright green color and are arranged in plane on the stem which gives the fronds a flat appearance
- Leaves are unique, natural curving, arching and twisting form
- Fruit is purplish red colour with 1.9 cm long, eaten by monkeys
- Suitable to grow from warm-temperate to tropical areas

76. *Phoenix sylvestris*

Common names: India Date, Khajuri, Silver date palm, Sugar date palm, Sugar palm, Sugar palm of India, Toddy palm, Wild date, Wild date palm

Description:

- It is an unbranched, erect, tall, dioecious, evergreen and slow growing nature
- It grows up to 4 to 8 metres in height with large persistent leaves in a terminal tuft
- Young trunk bear triangular shaped leaf that becomes more diamond-shaped with age
- On older trees, aerial roots tend to be present at the base of the trunk
- Leaves are pinnately compound and blue-green, and they can grow to 10 feet in length
- Leaflets can reach approximately 18 inches long and grow opposite to one another on the rachis in such a way that the entire leaf looks flat
- The petiole, or stem that attaches the leaf to the trunk, is 3 feet long and armed with spines
- Yellow inflorescences can reach length of 3 feet, are heavily branched, bear small white blossoms
- Fruit is oblong and of 1 inch long that occur in orange clusters, turning dark red to purple when mature
- The fruits are seedy and the seed occupies more than half of the fruit and of sweet nature
- It is an ornamental tree and can also be used as an avenue plant
- The trunk is used by the villagers in the construction of houses, it forming the supporting beam of the roof
- Halved trunks are used for diverting the water into the turbines of water-mills
- The leaves are used for making brooms, fans, floor mats, *etc*
- Flowers, borne on a spadix covered by a spathe which is 29.5 cm long; the spathe separates into two boat-shaped halves, exposing the flowers at maturity; both male and female inflorescences, about 25 cm long, bearing about 2,800 flowers
- The plants growing in the plains yield a good amount of juice which is used for making toddy and jaggery. The juice, as such, can also be drunk

- The tree provides a good fodder for milk cattle and is believed to increase the fat content of milk

- The fruit is cooling, oleaginous, cardiotonic, fattening, constipative, good in heart complaints, abdominal complaints, fevers, vomiting and loss of consciousness

- The juice obtained from the tree is considered to be a cooling beverage

- The roots are used to stop toothache

- The fruit pounded and mixed with almonds, quince seeds, pistachio nuts and sugar, form a restorative remedy

- The central tender part of the plant is used in gonorrhoea

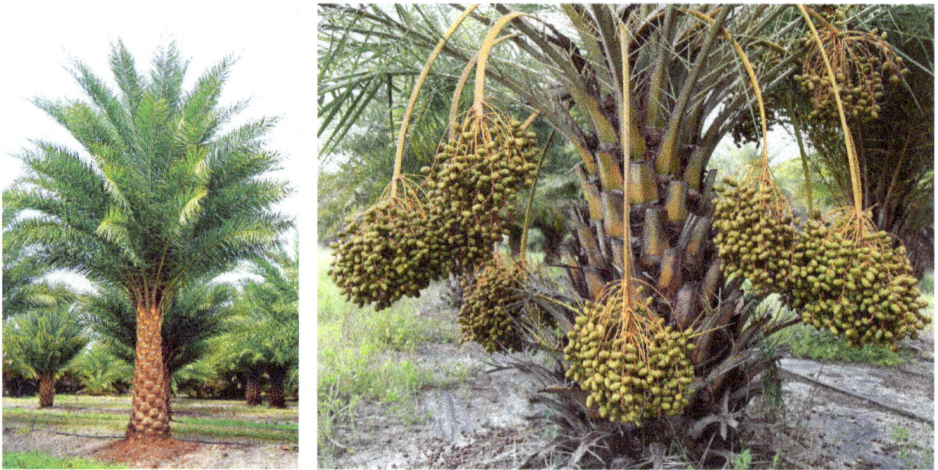

77. *Pinanga coronata*

Common names: Bunga, Ivory crown shaft palm, Pinang palm

Description:

- It is a small, clustered, undergrowth palm

- Trunk is erect, unbranched, (10) m tall, 1.5-7 (10) cm in diam., with internodes 4.5-12 (20) cm, scars 0.5-1.2 cm; stem surface green to brownish green

- Crownshaft swollen elongate, 50-100 cm long, 2.5-10 cm in diameter, slightly wider than the stem, green, yellowish or brownish green, or brownish to reddish yellow when adult, with brown scales, ligule poorly developed

- Leaves 4-7 in the crown; sharp near the apex; leaflets 6-30 on each side of rachis

- Inflorescence is pendulous or erect, green when young, becoming yellow pink to red
- Fruit is obovoid, ellipsoid to ovoid, 11-15 x 6-10 mm, green when young becoming yellow pink, red to brownish red

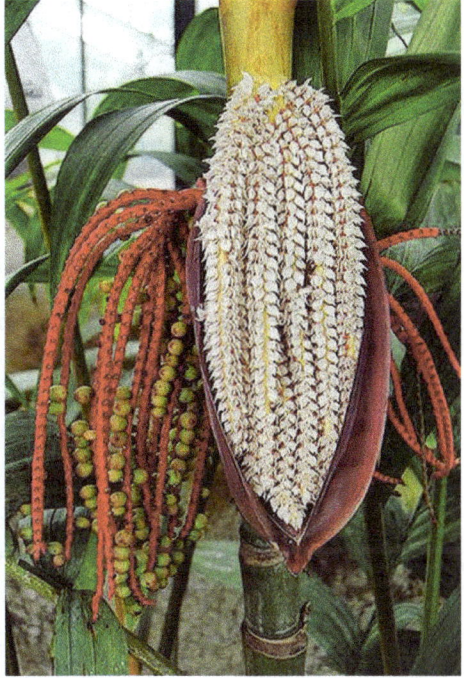

78. *Pritchardia pacifica*

Common names: Fan palm, Fiji fan palm, Pacific fan palm

Description:

- It is a palm that grows up to 10 m with a single stem that is up to 30 cm in diameter
- Leaves are fan shaped, 1.2 m wide and equally long, held on petioles of 1.2 m in length
- The large, flat and rounded leaves are divided 1/4-1/3 into many stiff-tipped segments
- The inflorescence is composed of 1-4 panicles which are shorter than or equaling the petioles in length
- Fruits are small, shiny dark brown to purplish black and spherical shape of 1.0 cm in diameter
- It is widely grown as an ornamental plant in tropics

79. *Pseudophoenix sargentii*

Common names: Buccaneer palm, Cherry palm, Florida cherry palm

Description:

- It is a very slow growing palm which reaches up to 10-15 ft tall and 10-15 ft wide
- It has a single smooth grey trunk that is ringed with scars of shed fronds
- The trunk has no crownshaft and topped with dark green pinnate shaped, feathery leaves of 15ft long
- Flowers are small, yellow colour that grow on a branched inflorescence
- Fruits are round, berry type and of bright red colour when ripe

80. *Ptychosperma elegans*

Common names: Alexander palm, Solitaire palm

Description:

- It is a single trunked tree that can reach 6.1–12.2 m in height
- The trunk is 2.5–10.2 cm in diameter and is light grey or almost white, with dark leaf base scars encircling the trunk
- Leaves are pinnately compound which reach the length of 1.8–2.4 m and are attached to a 0.30 m length petiole
- Leaves are dark green on the top and grey-green on the underside
- Inflorescences produce white male and female flowers
- Fruits are bright red, about one inch in diameter, and egg shaped
- It is popular as a house plant

81. *Ptychosperma macarthurii*

Common names: Hurricane palm, Macarthur feather palm, Macarthur palm

Description:

- It is an elegant palm, will eventually reach a height of 6 to 18 metres
- It is an evergreen clustering palm of variable height with a slow to moderate growth rate
- It possess multiple, slim and ringed grey trunks
- Trunk is topped with soft green, feathery, flat, broad leaves with tips that appear jagged and torn

- It produces off-white or whitish-green flowers
- Each fruit contains a seeds with five ridges along its length
- It can be planted singly or in groups
- It is popularly grown for its ornamental beauty
- Fruits will ripen to a brilliant red tone, almost roundish-shaped and about 1.3 cm long
- It prefers bright, indirect light but will also tolerate lower light levels
- It is an excellent palm to plant along walkways, near swimming pools *etc*
- It is excellent for landscaping in tropical areas
- Suitable in parks and gardens as an ornamental specimen or grown along sidewalks, road dividers, highways and byways
- It looks great to flank both sides of home entrances or driveways with these majestic-looking palms
- It is very adaptable to varying light and planting conditions
- It can be grown in containers and be ideal as indoor plants or to add interest at patios, decks, shopping malls and other public areas

82. *Raphia farinifera*

Common names: Madagascar raphia palm, Latrum palm

Description:

- It is monocarpic or hapaxanthic, flowering and fruiting only once, followed by death
- Trunk is sheathed with persistent leaf bases which grows up to 10 m tall and about 1 m in diameter
- Inflorescence is pendant and of 3 m length

- Flowering occurs when the tree is some 20-25 years old and it takes a further 5–6 years from flowering to ripe fruit, all fruits ripening together
- Fruit is oblong to ovoid, 5-10 cm in length, with imbricate, glossy, golden-brown scales
- It requires swampy conditions to thrive

83. *Ravenea rivularis*

Common name: Majesty palm

Description:

- It possesses single trunk with no crown shaft and of fast growing nature
- Leaves are upward-arching that are divided into long, thin fingers
- The trunk is broadest at it's base and tapering towards top
- It prefers full sun along the coast but does quite well in half day sun
- Leaves are of pinnate type, rather flat in cross section and six to eight feet long with a short bare petiole
- Mature fruit is red in color and seed is tan in color
- It can be grown as house plant
- It also prefers damp conditions to thrive well

84. *Reinhardtia gracilis*

Common names: Window palm, Window pane palm

Description:

- It is a small clumping palm, with thin canes grows up to 1.5m high
- The windowpane palm is called that because the leaves are fishtail shaped with usually only two broad leaflets per side and have small holes in them near their base
- Inflorescence produce oval fruits, 1 cm long, purple to black when ripe
- It prefers shade and moist areas like rainforests
- It is grown as a container plant
- It is suitable for both indoor and outdoor decoration

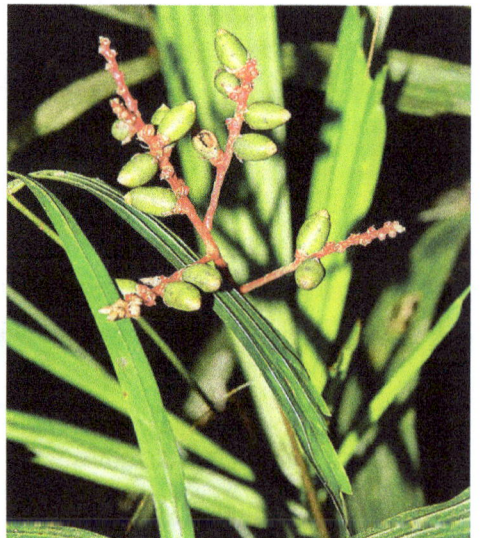

85. *Rhapidophyllum hystrix*

Common names: Blue palmetto, Creeping palmetto, Dwarf saw palmetto, Hedgehog palm, Needle palm, Porcupine palm, Spine palm, Vegetable porcupine

Description:

- It is a fan palm, leaves with a long petiole and of fast growing nature
- It is the second-hardiest palm known after *Trachycarpus fortune* (Chusan palm)
- Each leaf is up to 2 m long, with the leaflets up to 60–80 cm long
- It has a tightly compact inflorescence of 15-30 cm long and held close to the stem, barely peeping above the leaf bases
- Flowers are tiny yellow to purplish-brown colour and of dioecious nature
- The fruit is a brown, drupe about two cm long
- Seeds are red to brown and roughly spherical and are about 2.5 cm in diameter and have a fuzzy fleshy covering
- It prefers ample water, even marshy ground, full sun and shelter from wind

86. *Rhapis excelsa*

Common names: Bamboo palm, Fern rhapis, Ground rattan, Lady palm, Little lady palm, Slender lady palm, Miniature fan palm

Description:

- It is a delicate palm forms dense clumps of bamboo-like stalks topped with very dark green, broad, fan-shaped leaves
- It is of slow-growing, extremely versatile and long-lived
- It has big, thick leaves with blunt tips and wide segments
- It is a dioecious palm species produces a small inflorescence at the top of the plant with spirally-arranged, fleshy flowers containing three petals fused at the base
- Ripe fruits are small, fleshy, white, oval shaped and black when ripe
- It can be an effective "accent" in a shrub border or near an entryway
- Elegantly graceful, adorned with richly patterned leaves, and displaying exquisite beauty
- It can also be grown indoors, both in homes and offices
- It is tolerant to various types of climates including low light conditions
- It grows very well under dry and humid climates in tropics
- Propagated by seed and splitting the clump

87. *Rhapis humilis*

Common names: Reed rhapis, Slender lady palm

Description:

- It grows to a height of 6 meters tall with sheaths of 18–40 mm in diameter
- They form bamboo-like clumps and trunks are covered with a net of dark, light brown fibrous leaf sheaths and are topped with clusters of 5 to 10 leaves and the whole plant presents a finer appearance
- Leaves are fan-like and are dark green in colour, each leaf has around 15 leaflets
- It is of very slow growing nature, which makes them well suited for the home
- It can be used to create attractively dense screens and hedges
- Propagation is done by division
- Male flowers are produced on short flower stalks, no females are available under cultivation

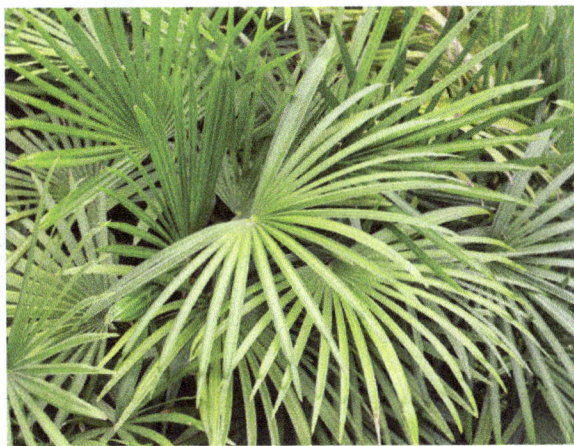

88. *Rhopalostylis sapida*

Common names: Brush palm, Feather duster palm, Nikau, Nikau palm, Shaving-brush palm

Description:

- It is a very slow growing palm which reach a height of 10m or more with a crown of spreading leaves that sheath to form a bulbous base

- It will take 15 years or more to form a trunk and is ringed by the scars left by the sheathing bases of the fallen leaves
- The leaf bases are large and encircle the trunk
- It can take 30 years or more to start flowering and fruiting
- Flowers are purple followed by brilliant-red hard berries which hang from just below the base of the leaves
- The berries take a year to ripen. The nectar and ripened red berries are food for native birds
- Leaves made a very effective waterproof roof and wall thatch and are woven into baskets and floor mats
- Fresh young shoots around the growing tip are cooked and eaten
- It looks most impressive when planted in groups

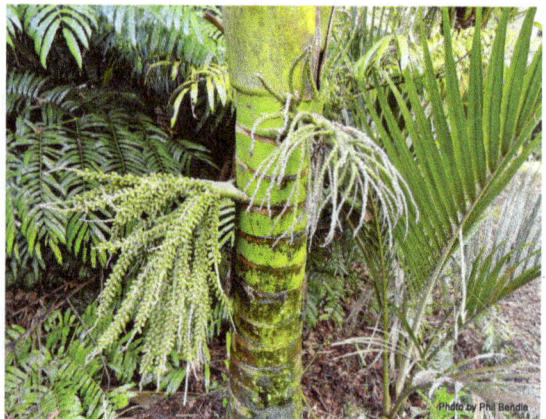

89. *Roystonea regia*

Common names: Cuban royal, Cuban royal palm, Florida royal palm, Royal palm

Description:

- It is a large palm which reaches a height of 20–30 m tall
- The trunk is stout, very smooth and grey-white in colour
- Leaves are 4 m long and have short petioles and color is green. Below the leaves, there is a long, smooth tubular crown shaft
- The flowers are white with pinkish anthers
- The fruits are spheroid to ellipsoid in shape, 9 - 15 mm long and 11 mm wide

- They are green when immature, turning red and eventually purplish-black as they ripe

- Propagated by seed. Seedlings grow fast and within 10 years produce a large trunk

- It is a good ornamental plant and is also used as a source of thatch, construction timber, and as a medicinal plant

- The fruit is eaten by birds and bats (which disperse the seeds) and fed to livestock

- They can be used as a centerpiece specimen plant

- They are wonderful along streets or long driveways

90. *Sabal minor*

Common names: Blue palmetto palm, Bush palmetto, Dwarf palm, Dwarf palmetto, Dwarf palmetto palm, Little blue stem, Scrub palmetto, Swamp palmetto

Description:

- It is a fan palm which grows to a height of 1 m height with a trunk of 30 cm diameter

- Usually stemless, the leaves are arising from an underground stock

- Leaf blades longer than the leaf stalks, 4 feet wide, dissected, the narrow segments notched at the tip

- It forms a trunk when grown in standing water. Each leaf is 1.5 - 2 m long, with 40 leaflets up to 80 cm long, conjoined over half of this length

- The flowers are yellowish-white, 5 mm across, produced in large compound panicles up to 2 m long, extending out beyond the leaves

- The fruit is a black drupe, 1 - 1.3 cm diameter in long clusters and their weight causes the flowering stalk to arch downward, sometimes to the ground

- Very hardy to cold conditions, grows very slowly in temperate regions

91. *Sabal palmetto*

Common names: Blue palmetto, Cabbage palm, Cabbage palmetto, Cabbage tree, Common Palmetto, Palmetto, Palmetto palm

Description:

- It is an evergreen, medium-sized, spineless palm that can reach up to 20 m in height

- Trunk is erect, unbranched which possess greyish to brownish bark with distinctive pineapple-like markings where the old leaf stalks were attached

- Leaves are fan like, stiff and palmately compound which spread in all directions as they emerge from the top of the trunk

- Flowers are abundant, small, fragrant, white coloured and are borne upon drooping, branched clusters

- The berry-like fruits are small (1.5cm), shiny and black, produced in abundance

- Each fruit contains one seed

- It is highly tolerant of salt spray and inundation by brackish water

- Trunk is used for wharf pilings, docks, and poles

- Brushes and whisk brooms are made from young leafstalk fibers

- Baskets and hats are made from the leaf blades

- It is an ornamental and street tree and one of the hardiest palm
- It is highly resistant to infection by pathogens
- It prefers poorly drained soils, and often grows at the edge of freshwater and brackish wetlands
- It can tolerate flooding
- It provides food for many birds and mammals

92. *Serenoa repens*

Common names: Saw palmetto, Scrub palmetto

Description:

- It is a small, hardy palm, growing to a maximum height of around 2–3 m
- Its trunk is sprawling, and it grows in clumps
- It is extremely slow growing, long lived, fan palm
- The leaves are light green inland, and silvery-white in coastal regions
- The flowers are yellowish-white, about 5 mm across, produced in dense compound panicles up to 60 cm long
- Fruit is a yellowish green in the unripe state, gradually turning blue-black as it ripens and is an important food source for wildlife and historically for humans
- The seeds are collected in vast numbers for use in medicine
- It is especially resistant to fire even though its foliage is highly flammable
- It recovers from fire very quickly, with cover returning to pre-burn levels within a year

- The drug made from this palm has been successfully used to treat, prostrate swelling, bladder infections and urinary tract infections

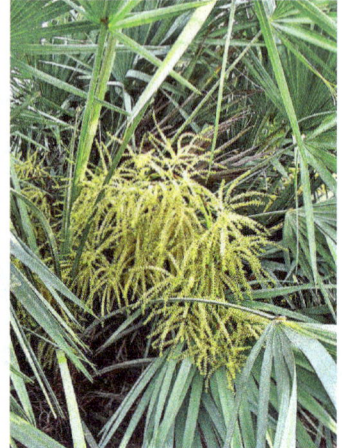

93. *Syagrus romanzoffiana*

Common names: Giriba palm, Queen palm

Description:

- It is a single-trunked, medium sized, fast growing palm reaching a height of about 10-15 meters tall
- It has smooth straight grey trunk ringed with evenly spaced leaf scars and topped with a large canopy of pinnate leaves forming a graceful, drooping canopy
- The dead fronds are persistent and often require pruning
- The ornamental, bright orange dates are produced in hanging clusters and ripen during the winter months and are edible
- It is most suited for acidic and well-drained soils
- It is drought resistant and tolerant to wide range of conditions
- It is used for parkway plantings
- The trunks are frequently hollowed out to make water pipes
- The leaves, or the fibers obtained from them, are used for making baskets and hats
- Its leaves and inflorescences are used as cattle fodder, especially for milking cows
- Fruits are bright orange and fleshy, round and an inch in diameter

- It is used for various constructional purposes, as stepping boards over swampy areas, footbridges and rustic piers in salt water
- The wood is moderately heavy, hard, very durable in salt water

94. *Thrinax morrisii*

Common names: Brittle thatch, Brittle thatch palm, Broom palm, Buffalo-thatch, Buffalo top, Peaberry palm, Pimetta, Key palm, Key thatch, Silver thatch palm

Description

- It grows slowly and reaching a height of about 20 to 30 feet tall
- Trunk is smooth, slender topped with fronds
- The fronds are of shimmering silver/white underneath
- Flowers are small and in white colour
- Fruits are small, round and white colour
- It is a best container palm and small enough to be popular in residential landscapes
- It is often planted as a single specimen or in groups of three to 'accent' an area
- Due to the coarse texture, they make a nice entryway palm to attract attention to the front door of a building
- It should be grown in full sun or partial shade and is highly drought and salt tolerant
- It is an ideal palm for seaside applications
- Leaves are used for thatch making

95. *Thrinax parviflora*

Common names: Broom palm, Florida thatch palm, Iron thatch, Mountain thatch palm, Jamaica thatch palm, Palmetto thatch, Thatch, Thatch pole

Description

- It is an evergreen, single-stemmed palm tree grows to a height of about 10-13 m tall
- Trunk is slender, 5 - 15cm in diameter topped by an open crown of small, very thick and leathery fan leaves
- Leaves are curiously twisted and curled, heavily veined grass green segments
- It thrives in subtropical climates
- It is tolerant to coastal exposure
- At low elevations tend to have short, thick trunks, whilst at higher elevations they are taller and more slender

96. *Thrinax radiata*

Common names: Florida thatch palm, Jamaica thatch, Sea thatch, Silk-top thatch

Description

- It is a medium sized, slow growing; solitary palm can grow up to 12 m tall
- Trunk is slender with 13 cm in diameter with obscure, incomplete leaf scars
- It has no crown shaft and the canopy appears to emerge directly from the trunk
- The leaves are palmate and divide into segments about halfway down their length with 4 to 5 feet wide and 2.5 feet long with drooping leaf tips
- The upper leaf surface is dull green; the undersurface is light yellowish green
- Fruits are small (7-8 mm in diameter), spherical and white when ripe and is inedible
- This palm does best in full sun but can tolerate moderate amounts of shade
- It is a great palm for coastal regions
- It is grown in containers and also in arboretums
- Leaves are used for roof thatching, roofing, broom construction, handicrafts, and food wrapping
- Trunk have recently been used to construct lobster traps by fishermen

97. *Trachycarpus fortunei*

Common names: Chinese windmill palm, Chusan palm, Fan palm, Hemp palm, Windmill palm

Description

- It is a fan palm possess single stem, grows to a height of about 12–20 m tall

- The trunk is very rough with the persistent leaf bases clasping the stem as layers of coarse fibrous material

- The flowers are dioecious nature and produce yellow (male) and greenish (female) flowers in large branched panicles up to 1 metre long

- The fruit is yellow to blue-black, reniform (kidney-shaped) drupe

- It does not grow well in very hot climates

- It is grown as an indoor plant when young and of slow growing nature

- It is moderately salt tolerant and can be planted behind the first line dunes or against a structure that will shield it from direct exposure to sea breezes

- It is a good pot plant for patio, deck or pool

- One of the most cold tolerant palms in the world

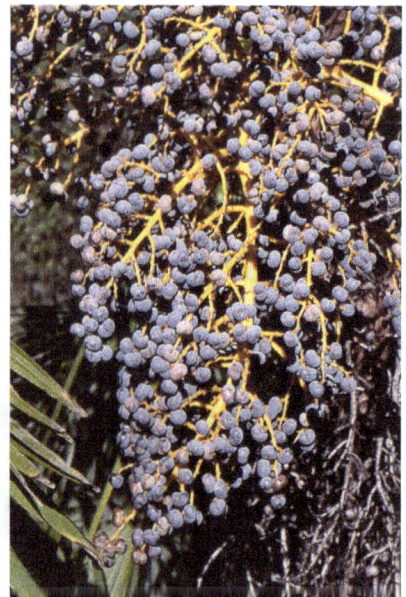

98. *Trithrinax acanthacoma*

Common names: Buriti palm, Spiny fibre palm, Brazilian needle palm

Description

- It is very attractive, medium sized, solitary palmate palm, with an unusual woven fibre wrapping the trunk, which includes the old spines

- It can grow to a height of 12 m in areas that are subject to fire and grazing tends to have bare trunks

- Leaves are green with contrasting glaucous waxy backs, split around half way

- Leaf tips being split into something resembling a snake tongue with two shallow forks each tipped with a small spine

- It bears large bunches of white/pale green oval fruits about 1.5 cm long

- It prefers sunny, moist, but well drained and slightly alkaline soil

- It is of drought and frost tolerant, spines on the trunk soon become apparent even in young plants

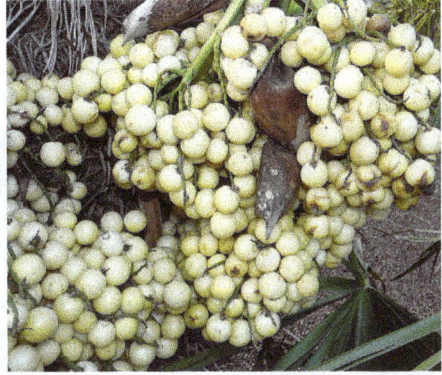

99. *Washingtonia filifera*

Common names: American cotton palm, Californian cotton palm, Californian fan palm, Californian palm, Cotton palm, Desert palm, Desert fan palm, Desert Washingtonia, Fan palm, Northern Washingtonia, Petticoat palm

Description

- It is an evergreen; fan shaped palm can grow up to 18.3 m tall with a crown spread of 4.6 m, one of the most popular palms in the world

- The massive grey trunk is barrel shaped and ringed with old leaf scars, and may reach over 1.0 m in diameter at its widest point

- Leaves are grey-green, palmate, each about 0.9-1.8 m long. The petioles of mature palms are armed along the margins with curved thorns

- Young palms are largely unarmed

- The individual leaflets are pendulous and swing freely in the wind

- Abundant cotton-like threads on and between the leaflets persist even when the palm is mature, they hang down, forming a skirt, cover the entire length of the trunk

- They have long thread-like white fibers and the petioles are pure green with yellow edges and *filifera*-filaments, between the segments

- If old leaves are not removed, they form a continuous "petticoat" from the crown all the way to the ground

- It produces numerous branching flower clusters that project out and often hang downward from the leaf crown

- The bisexual blossoms are white and yellow and give rise to oblong or round red-black fruit, each about 1.3 cm in diameter

- Flower stalks produce abundant small round fruits black, when ripe

- The fruits contain a single seed, approximately 0.6 cm in diameter

100. *Washingtonia robusta*

Common names:　Mexican Washington palm, Mexican Washingtonia, Mexican fan palm, Skyduster Southern Washingtonia, Thread palm

Description

- It grows faster and will reach a height of about 25 m tall
- Trunk is grey in colour, ringed with closely set leaf scars although usually at least part of the trunk remains covered with dead leaves that hang in a thatch
- The solitary trunk, about 25-30 cm in diameter, bulges at the ground and becomes slender as it approaches a crown of large palmate leaves with gracefully drooping leaflet tips
- Leaves are palmate and leaflets are up to 1 m long
- As the leaves die, they fall against the trunk to create the "hula skirt" effect
- Inflorescence is 3 m long, with small pale orange –pink flowers
- Fruit is a spherical, blue-black drupe, 6–8 mm diameter; edible, thin-fleshed
- It is planted along streets, in parks, and next to apartment buildings

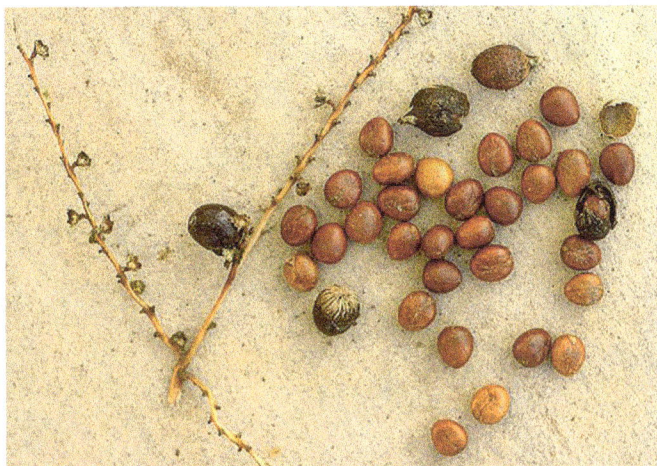

101. *Wodyetia bifurcata*

Common names: Foxtail palm

Description

- It is thin trunked, pinnate palm which gives very attractive appearance and makes a superb landscape effect
- Trunk is smooth, slim and growing to a height of 12 m
- Leaves are very plumose with leaflets coming off the leaf stem at all angles

- Leaf color is green to dull green and the crown shaft is silver to green
- Trunks may bulge slightly at their base and in their mid-section
- Flowers emerge right below the crown shaft, branched and tan or white in color
- Fruits are 2 inches long. Olive green to green in the early stages and become orange red when ripe
- It prefers sunny condition, tolerates heat and grows well in subtropical to tropical regions
- It's not suitable to grow as a container plant

7

Germination of Palm Seeds

Palms are commercially propagated by seeds. Basic requirements for the seed germination of most of the palms are fresh seed, good sanitation, proper medium, proper hydration and adequate heat.

Fresh seed

To check the freshness of the seeds, cut open a sample seed and inspect the endosperm and embryo. The embryo should be fresh, firm, and not discolored. If the interior of the seed is rotten or has an unpleasant odour, it is unlikely to germinate. The endosperm is of two types, they are homogeneous or ruminate, and may be hard, oily, or even hollow. If the inside of a homogeneous seed is off-color, such as brown or gray, or if it smells bad, the seed is old or was harvested before maturity. Such seeds are also unlikely to germinate. In a ruminate seed, the seed coat is infolded, creating dark, tangled streaks in the endosperm. Ruminate seed is more difficult to assess because of its more complex appearance.

Removing the fruit pulp

The fleshy or fibrous fruit pulp frequently contains growth inhibitors. Removing it before planting will improve the germination percentage. But it varies with the quantity and type of seeds, but most begin with a preliminary 48-72 hours soaking in water. Soaking causes the pulp to ferment, which weakens it for easier removal. Change the water daily during the soak. Fruit that is slightly immature should be placed in a tightly closed plastic bag and keep it in a warm spot for a week. This promotes ripening and softens the outer flesh for easy cleaning. Sometimes the seeds need to be soaked further to soften the pulp, but it is not same for all palm seeds.

There are several ways to remove the seed coat. With small quantities of seeds, simply rub them by hand against a fine-meshed screen and wash away the pulp with water. Another way is shaking a small amount of seed in a closed container with water by hand and also by small, rough-edged rocks. Pour off the water and pulp occasionally, add more water and shake again, until the seeds are completely clean. Seeds can also be cleaned with a knife or other sharp tool, but this is slow and a little dangerous.

Recently several motorized cleaning devices are available in the market and helps to do the cleaning work in a short time. This kind of devices is highly useful for commercial large-scale growers. For smaller quantities, usage of rock tumbler holds good. Put rocks and water inside the tumbler with the seeds. Larger seed-cleaning machines can be purchased or fabricated. Some large-scale growers and seed dealers may use cement mixers for this cleaning process. The seeds are rotated in the drum for 10-45 minutes with water and rough-edged rocks of 7-10 cm. The time will vary with the machine and the type of seed and rocks. Some seeds are brittle, and without proper care, it may be damaged by proper cleaning. Among large-seeded palms, *Actinorhytis* is particularly brittle and prone to damage, and many smaller seeds, such as *Pinanga*, must also be handled with care. When cleaning seeds, remember that the flesh of some types contain crystals of calcium oxalate, a skin irritant that can cause severe pain on contact, depending on the individual's sensitivity. For this reason, *Ptychosperma, Arenga, Caryota,* and *Wallichia* should be handled with care.

Sanitation

Damaging insects such as seed-boring beetles may come with seeds. They may reduce germination and can easily spread to other seed batches. To minimize these risks, immediately after collecting the seeds, they should be soaked in a contact insecticide solution once the fruit pulp has been removed. Soak the small, thinner-shelled seed, such as *Pinanga*, for 15 minutes. Soak larger, harder and less permeable seeds (*Mauritia flexuosa, Bismarckia nobilis, Parajubaea cocoides,* and *Jubaea*) for 20 to 45 minutes. After the insecticide soaking, rinse the seeds in clean water for 20 minutes.

After cleaning the seeds, hydrate them by soaking them in water for 24 hours, especially if the seeds are not soaked in water while collecting them from the pulp.

Within 24 hours most fresh, viable seeds will sink. There are exceptions such as *Manicaria saccifera* and *Metroxylon vitiense*, whose viable seed will float even after cleaning and soaking.

If the collection of certain palm seeds is very rare and getting infested, don't discard a batch of heavily infested, damaged seeds. For very rare type palm seeds, even a single seed germination could be valuable, plant it. Fungi flourish in the heat and humidity necessary for good germination, so equipment, fixtures, seeds and growing medium must be kept clean to prevent damping-off and other disease problems. It is better to soak seeds in a fungicide solution before planting.

Planting medium

Germinate the easily germinating palm species in a commercial mix of peat moss or sterile sphagnum moss mixed with an equal amount of perlite or vermiculite. It is also good to use coir compost in place of peat moss or sterile sphagnum moss in the above mixture. Sand, wood chips, screened rock or volcanic cinder screened to a maximum size of 9 mm which act as a substitute for vermiculite or perlite. The medium should be very porous and drain extremely well. All containers should have plenty of holes in the bottom to ensure quick and thorough drainage.

When containers and planting medium are ready, lay out the seeds on the surface, and before covering them, dust with a commercial insecticide. Bury the medium to a depth of half the seed diameter and then cover everything with finely screened sand (3-6 mm particle size), thick enough so that it will not wash away during watering. This top-dressing dries out quickly and discourages the moss that grows on peat. When planting is completed, place the containers on clean benches, 60-90 cm above the ground. Be sure to label the containers with a waterproof and fade-proof marker.

Palm seeds, known as remote germinators may require special treatment and a little extra patience. It pushes a shoot downwards as much as 20-25 cm before sending up the first leaf. The larger ones such as *Voanioala* and *Borassodendron*, should be planted in deep containers such as large tubs, or be transferred to such containers soon after germination. If seeds are large enough, it may be planted directly in the ground.

Easy germinators	Difficult germinators
• *Dypsis decaryi*	• *Basselinia* species
• *Pinanga kuhlii*	• *Parajubaea cocoides*
• *Pinanga crassipes*	• *Neoveitchia storckii*
• *Archontophoenix alexandrae*	• *Jubaeopsis caffra*

Easy germinators	Difficult germinators
• *Chamaedorea elegans*	• *Jubaea chilensis*
• *Chambeyronia macrocarpa*	• *Lavoixia macrocarpa*
• *Licuala grandis*	• *Physokentia insolita*
• *Veitchia joannis*	• *Pseudophoenix* species
• *Washingtonia filifera*	• *Rhopalostylis baueri*
	• *Howea fosteriana*
	• *Voanioala gerardii*

Hydration and heat

One of the most important factors in seed germination is proper hydration, followed by constant high heat. Maintaining proper hydration is most important while comparing the other one. Water the containers thoroughly, but just as important, let them dry out thoroughly before watering again. Over-hydration can drastically reduce the germination percentage. Once seeds begin to germinate, the containers will require more frequent watering. Seeds should be kept at 26-35°C. Sometimes, constant bottom heat may be provided by means of electric pads on their benches.

Plastic bag method

For the palm seeds which is of rare availability and for the seeds which finds difficulty in germination, the most reliable method of germination is the Plastic Bag Method. For this method, seeds are blanketed in damp sphagnum moss and germinated in zipper-type, re-sealable plastic bags. Thoroughly saturate the sphagnum moss with water and check it until no more water can be expressed. Place the seeds and the sphagnum moss inside the plastic bags (along with a label) and keep the bags at 26-35°C. Check inside the bags periodically to ensure that the sphagnum moss has not dried out. Once seeds have germinated, place them in separate pots containing the potting mix as given above and then follow the same process. When transferring the germinated seeds to seperate pots, it has to be given a fungicide drench. Germination setups can also be improvised from plastic foam boxes with tight fitting lids, such as are used to pack fish or fruit. Fill the boxes 1/3 full of fine perlite pieces and lay the seeds on top. Use a hand mister to dampen thoroughly the seeds and perlite, replace the lid and place the box in a warm location. These germination boxes are space-savers, because they can be stacked. The tight-fitting lids help to keep out fungus and insects, but the boxes should be checked periodically for hydration and germination.

8

Value Added Palm Based Products

Though value added products are prepared from several palm species, coconut and palmyrah based value added products gained importance in market. Coir based handicrafts from coconut are fetching higher price in the international market. Hence, this chapter mostly deals with details of value added products made out of coconut and palmyrah palms.

1. ***Aiphanes aculeata* (syn: *A. caryotifolia*)** - Seeds are used to make candles

2. ***Aiphanes erosa*** - The endosperm of the seeds is edible, and is similar in taste to that of a coconut

3. ***Borassus flabellifer***

Edible products	Neera, Palm Jaggery, Palm Candy, Palm Sugar, Fruit Jam, Syrup, Chocolates, Toffees and Confectionery items made out of Palm Sugar
Non-Edible products	Palm fibre brush, Palm leaf fancy articles, Palm leaf visiting cards.

I. EDIBLE PRODUCTS

a. **Neera**: Neera is the top most economic product of palmyra. It is good source of minerals like calcium, phosphorus and iron. Vitamin A, citric acid, niacin, thiamine and riboflavin are present in neera. Neera acts as laxative and diuretic.

b. **Endosperm (Nungu):** Jelly like endosperm of young fruit of 60 – 70 days old is called nungu which is a summer delicacy. It is very nutritive.

c. **Tuber (Apocolon):** Mature tuber is brittle and breaks off easily which is a rich source of carbohydrates. Optimum time for harvesting of tuber is 135 days after sowing.

d. **Fruit:** Fruit gives sweet aroma with fleshy mesocarp. Fruit is roasted and consumed or mixed with flour and baked into flakes.

e. **Thavan (spongy haustorium):** Thavan formed during germination of seed nut is spongy, sweet and delicacy.

f. **Palm jaggery:** It is also called as palm gur. Jaggery is made by boiling neera in a galvanized iron pan at 110°C. Neera gets transformed into viscous fluid which is poured into shells and allowed to harden. About 8 liters of neera is required to get 1 kg of jaggery. Jaggery contains good nutritive and medicinal values. Major problem of jaggery storage is blackening of colour in short period which needs to be corrected.

g. **Palm candy:** Neera free from debris boiled in an alloy vessel adding small quantity of superphosphate. After uniform boiling, the liquid is allowed to cool. After removal of sediments, it is heated to 110°C for 2 hours until it reaches honey like consistency. The fluid is then allowed to cool and poured into crystallizer. Sugar crystals start forming after 45-60 days.

h. **Palm honey:** Neera is heated for 2 hours to obtain the honey like consistency. The syrup is then transferred to mud pots. Ripe, dry and shelled tamarind fruits devoid of seeds are added into the syrup. About 1 kg of fruit is required for adding to 10 liters of syrup. The pot is closed tightly with cloth and vessels are kept in a shock proof, cool and dry place for 130 – 180 days. Sugar crystalises on the sutures of tamarind and the fruits become delicious.

Palmyrah Sap

The male inflorescence is extensive and branches into long thin finger like processes which bear many male flowers. The female inflorescence axis is thick and robust and bears 5 to 15 fruits depending mostly on its nutritional status.

Male inflorescence is a compound spadix, main axis which has a number of spathes, from whose axis originate 2 – 6 branches, each bears 2 - 3 club shaped spadices of male flowers. Each spadix has a large number of brown flattened, squama like bracts, arranged spirally, each protecting a cluster of 10 – 12 male flowers. Opened male flowers appear on the surface of the spadix and they drop off after the dehiscence of anthers. Each male flower has 6 tepals arranged in two whrols with six stamens and a rudimentary ovary in the centre. Female inflorescence is compound with a main axis having 2 - 6 branches. Each branch bearing a spadix consists of 3 - 12 female flowers. Female flower has 6 tepals in two whorls. Staminodes 6, small connate at the base forming a ring likes structure at the base. Centre is a tricarpellary ovary.

Tapping the sap from inflorescence

The tapper goes up the tree mostly twice a day (morning and evening) to prepare the young inflorescence tissues. Both male and female are tenderized by gentle massaging with a massaging horn and are then sliced with a sharp tapping knife The slices are less than a millimeter in thickness. Thinner slices extend the tapping life of the inflorescence. Slicing opens up the (xylem - phloem) channels for the sap to flow. During the course of the day, sap flow decreases as the freshly incised area turns brown, covered up by tannins. Hence in about 12 hours time, tapping has to be repeated to increase the flow of sap. The volume of sap collected per tree varies from 3 .2 l/day from the mature inflorescence of male trees to 8 l/day from the mature inflorescence of female tree. The total sugar content of the sap varies from 10.9 to 15.2 %, w/v. Inhibition of the oxidation of phenols to quinones by polyphenol oxidase by the application of EDTA (Ethylene Diamine Tetra Acetate) to the sliced end increases sap flow.

Sap Based Products

The fresh sap flowing from the inflorescence is sterile and if collected in a sterile container, no changes due to contamination by microorganisms can occur. This sap is a sweet clear watery liquid and contains sugars, vitamins and minerals. The pH of the sap is about 6.8.

1. Jaggery

Jaggery is the most important product made out of pathaneer. It is made directly by concentrating the pathaneer to a thick consistency. The sap is clarified by first straining through a mesh or cloth to remove suspended parts and where lime is used, the suspended calcium is removed through precipitation as calcium phosphate by adding a saturated solution of super phosphate until the pH is brought down to about 6.8. The delimed filtrate is boiled with frequent stirring until the right consistency is reached. The strike temperature is about 116°C. The concentrated syrup is poured into molds to set of 10 kg of jaggery from 72 litres of pathaneer.

Pathaneer is concentrated without deliming and the product is generally light brown (almost yellow) in colour. The product from Thailand is rather soft, unlike the products from India and Sri Lanka. The pathaneer is dehydrated slowly by heating over cinders with constant. Powdered palmyrah jaggery stirring. When it begins to solidify, it is stirred well to break it up into finer particles. The moisture content of the powered form is much lower than the solid form and it is packed in quantities of 500g and 1kg, in polythene bags. Analyzed jaggery produced in Sri Lanka and reported the composition of palmyrah jaggery as sugar (sucrose) 77.4%, moisture 10.83%, reducing sugar 0.51% and ash +insolubles 11.24%. Chemical analysis on jaggery from palmyrah has shown the presence of vitamins such as riboflavin (402mg/1008), vitamin B12 (15 mg/100), vitamin C, Thiamine and nicotinic acid. One of the problems encountered with palmyrah jaggery is the caustic taste due to residual $Ca(OH)_2$ and the poor keeping quality. The jaggery tends to turn into molasses on storage. The former problem can be solved if lime can be avoided, where several alternatives are available. As known to be hygroscopic, the psychrometric conditions of the atmosphere in contact with the jaggery are largely responsible for its conversion into molasses. Scientific and Industrial Research showed that they could store well if the relative humidity of the atmosphere is kept below 62%.

In India, the practice is to pack the semi spherical pieces of jaggery (size of half a coconut shell) in boxes of 10kg. The boxes are made out of mature palmyrah leaves. These boxes are stacked on wooden racks covered with a layer of hay or paddy straw. These procedures help to absorb the moisture and maintain a low relative humidity. In some store rooms, electrically operated dehumidifiers are installed to maintain a low relative humidity.

A spicy jaggery called 'puttu panagkatty' is made in certain parts of Tamilnadu. Here spices like cardamom, ginger, green gram *etc.* are added to the concentrated syrup just before pouring into the mould. Spicy jaggery is made as smaller semi spherical pieces packed in small boxes (250g) made out of palmyrah leaf. The 'Kuttan panagkatty' is sold at a higher price than ordinary jaggery. In jaffna the practice is to prepare the Jaggery in cylindrical moulds of different sizes woven out of palmyrah leaf. But recently the palmyrah Development Board of Sri Lanka has started producing jaggery in the form of 2 inch cubes packed in polythene bags.

Production of jaggery is still at a cottage industry level in all palmyrah jaggery producing countries. Tamilnadu jaggery production is under the Khadi and Village Industries Board, which manage 1149 co-operative societies, 5 district federations and one state federation, all of which are involved in the production of palmyrah Gur.

2. Treacle

Pathaneer when concentrated by heating to about 1/6th of its original volume forms thick dark syrup called treacle. The temperature should not rise beyond 107°C.

The syrup with a brix of 65% is ready to be removed from the flame, cooled and bottled. In Sri Lanka it is marketed in 750ml bottles. Usually the temperature is not controlled during the preparation and the sugar undergoes caramelisation resulting in a dark product with a disagreeable flavour. Improvements can be made in this product and syrup resembling 'golden syrup' can be prepared. Pathaneer collected in plastic containers without the addition of lime when concentrated under controlled temperature gave a much better product, similar to 'golden syrup'. The sugar content of the treacle is about 65%.

3. Sugar candy

Another important product of pathaneer in India and Sri Lanka is sugar candy. In the preparation of sugar candy pathaneer is boiled to 108°C and the brix raised to 67%. The concentrated liquid syrup is then allowed to crystallize slowly in U shaped metal crystallizers. The crystallizers used in India can hold about 30kg of syrup equivalent to 160 litres of pathaneer. In the crystallizer, strings are tied crisscross, so that the crystals can form around the string. The crystallizer is filled with the syrup, covered with a lid and the whole container is buried in a layer of paddy husk raised up to the brim of the U shaped crystallizer. This allows the crystals to be formed without any disturbance. The crystallizers are kept in this manner for a month and during this period small and large crystals are formed on the strings. At the end of one month, the crystals along with the string are taken out, washed by spraying water while it is being centrifuged in a bucket centrifuge. The washed crystals are allowed to sun dry graded according to their size and packed in polythene bags. In Sri Lanka sugar candy is prepared on a very small scale at Singainagar in the Jaffna district. The sugar candy price is actually three times the value of ordinary sugar (sugar cane) but yet the people buy because of its medicinal value. Sugar candy is used in indigenous medicine and it is believed to have curative values for reducing fevers, cures sore throat, dry cough, shot eyes due to heat, and urinary complications of pregnant women.

4. Pathaneer or Sweet toddy

Fresh unfermented sap is referred to as pathaneer, which is consumed immediately as a refreshing non-alcoholic drink or it is concentrated by heating to produce treacle, jaggery and other products. Pathaneer as a fresh drink is quite popular in Northern Sri Lanka. In folk medicine, sap is prized as a tonic, diuretic, stimulant, laxative, antiphlegmatic and amebicide. The pathaneer was light yellow in colour, usually kept in bottles. The vendor serves the pathaneer in glasses with ice cubes or in polythene bags with a straw for take away. In many parts of Indonesia, pathaneer is usually consumed fresh in the villages. In one of the main pathaneer producing villages in Surabaya around 20 small shops were found selling only pathaneer and young kernel of palmyrah during the season. They store the pathaneer in 5 to 10 litre white plastic cans and serve to the customers fresh usually with ice. Pathaneer is marketed through several marketing outlets, some which have cooling or chilling units to store the pathaneer for a longer period. Pathaneer is also bottled in 220 ml quantities. Bottling and cooling of pathaneer before sale, practiced in India is something the other countries could adopt.

5. Palmyrah sugar

At present India is the only country that produces granulated sugar from palmyrah pathaneer. Palmyrah sugar has been made in Sri Lanka on a very negligible scale, mostly on trial basis and at present there is no sugar production centre in operation. Attempts have been made to restart processing sugar, and to couple this industry with the more lucrative alcohol industry making use of the by—product molasses and convert into alcohol.

6. Palmyrah vinegar

Palmyrah vinegar has an ample amount of carotenes, vitamins (especially rich in Vitamin -B) and minerals besides the carbohydrates and proteins. Among the many health & medicinal values that linked to vinegar generally, the Natural Palmyrah Vinegar has its own class of values such as, 1. Cure for bladder infections 2. Antiseptic agent for cuts & abrasions 3. Relief for headaches and pains 4. Improving memory 5. Reduction of Urinary tract infections 6. Treatment for dandruff 7. Softening the skin

7. Palmyrah 'Toddy'

Palmyrah toddy is the natural fermented sap and contains about 5-6% (w/v) ethanol. During tapping of the sap, if the pot is not layered with slaked lime, the sugar (l 2. - l6 %, w / v) gets fermented to alcohol by wild yeast from the atmosphere. The sap containing sugar, vitamins and minerals forms a good culture medium for the microorganisms to grow. The microorganisms usually seen in Palmyrah toddy are yeast (mostly *Saccharomyces cerevisiae, S. chevalieri, Kloeckere apiculata, Schizosaccharomyces pombe*) and bacteria (*Bacillus cereus, B. sphericus, B. firmus*). The microbial count is maintained at high level by pouring off the clear supernatant liquid without disturbing the microorganisms at the bottom of the pot. The white color of the toddy increases in intensity with increasing microorganisms.

8. Palmyrah arrack

During the palmyrah toddy season from January to July each year, the excess toddy is collected from the taverns for distillation in one of four distilleries. The Thikkam distillery has a patent still with a capacity of distilling 1350 l (3 00 gallons) of toddy per hour while the other three centers can distill only 3150 l (700 gallons) of toddy per day. The alcohol content of the distillate from patent still is 86 — 90 % (w/ w). There is an excess of toddy during the season and to overcome this problem, low wine distillation was introduced. The low wine produced at the two centers is collected and distilled in the patent still during lean periods. However much toddy is thrown away in the local taverns as they cannot be processed further into bottled toddy or distilled. This has led to the rationing or limiting the quantity of toddy that a tapper can supply to the tavern per day during the season and also led to underutilization of skilled manpower and palmyrah palms.

9. Bottled palmyrah toddy

The toddy is filtered, bottled and pasteurized at 75"C with benzoate or metabisulphite being added as preservative. The bottled toddy is light opaque in color and retains its characteristic slight bitter taste and acidic smell. The bottled pasteurized toddy keeps for about 6 months without spoilage. Bottling of palmyrah toddy was started in 1988 by Palm Societies and three bottling centers were established in Jaffna, Sri Lanka. ln 1988, 20,000 gallons of toddy were bottled and by 1991. 40.000 gallons were bottled indicating its increasing demand.

Food Based Products

1. Palmyrah palakaram - (Kevum)

This is being prepared by using palmyrah pulp, wheat flour, baking powder, oil with little cane sugar and flavouring.

2. Palmyrah chocolate

This is prepared using palmyrah pulp, cane sugar, cocoa powder, fat and flavouring.

3. Palmyrah jaggery cake

This is prepared using egg, butter, semolina, jaggery, wheat flour, fruits nuts and flavouring.

4. Palmyrah thattu vadai

Palmyrah thattu vadai is prepared by using the ingredients like black gram, wheat flour, chilly piece, salt, turmeric powder, oil and flavouring.

Palmyrah tuber

The tuber as described previously is an important source of starchy food in the villages of Jaffna, Sri Lanka. In most of the other countries like India, Thailand and Indonesia tubers are rarely used. It was observed that in Thailand even the agriculturists and others who are involved with the palms were unaware of the tuber and its uses. The tubers were dug up from the soil and eaten after frying or cooking.

Freshly harvested tubers, after cleaning of the outer sheath and basal roots are boiled for about half an hour and the cooked tubers are eaten. It tastes like any other starchy tubers. It is also believed to have medicinal properties to relieve biliousness, dysentery and gonorrhea. The root of the seedling (tuber) is a diuretic, anthelmintic and a decoction is given for certain respiratory illnesses. In some parts of Nigeria, villagers eat the tuber with a belief that it is an aphrodisiac. But usually the tuber is converted into other suitable forms, which can be stored for further period to be used during the off-season.

Tuber Based Products

1.   Boiled tuber 'Pullukodiyal'

Boiled tuber are split into two halves and sun dried. This sun dried product is called "Pullu Kodiyal' which is eaten as a snack.

2.   Dried raw tuber flour & Seeval odiyal

Raw tubers are split into two halves. Usually they are stored in the form of dried pieces and milled only when needed for make odiyal flour.

3.   Boiled odiyal flour

Boiled tubers are cut into thin sections crosswise and sun dried. This is a slight variation of the (1st) and eaten direct. Boiled and sun dried 'Pullu kodiyal' is milled into flour (Boiled odiyal flour). Odiyal flour raw tubers are sun dried and milled into flour (Odiyal flour).

4.   Palm posha

Composition of instant palm posha products (rice flour black gram flour green gram flour soya bean flour boiled odiyal flour) .

5. Odiyal kool

The recipe for instant odiyal kool based on the traditional odiyal kool was developed by Palmyrah Development Board. The instant odiyal kool is popular among urban population. The dried debittered odiyal flour is mixed with dried parboiled vegetables such as string beans, plantain, manioc, potatoes, jak seeds, and green leaves. Dried powered prawns and fish are often added. These are packed with chilly and turmeric powder.

Edible Products From Palmyrah

Food based products

Fruit based products

Sap based products

II. NON-EDIBLE PRODUCTS

Leaf based products

Leaf lamina

Leaves of all ages from the ivory coloured young leaves to the brown coloured dead leaves of palmyrah are used. In the past, the thin whitish young leaves were used as paper for writing notes, memoirs, letters *etc.* The dead leaves with the lamina, petiole and the basal sheath are always used as domestic fuel in the villages. The lamina region is used to manure paddy fields in some regions.

Leaves of all ages from the ivory coloured young leaves to the brown coloured dead leaves of palmyrah are used. In the past the thin whitish young leaves were used as paper for writing notes, memoirs, letters *etc.* The dead leaves with the lamina, petiole and the basal sheath are always used as domestic fuel in the villages. The lamina region is used to manure paddy fields in some regions.

The mature green leaves, cut down from the palm are used for thatching and for fencing. The leaves have to be dried and the lamina stretched and flattened out before they could be used for the above purposes.

Usually the cut leaves were allowed to air dry for a day or two. Then the leaves are trampled with the feet to stretch the fronds and arranged in a line on the ground in two close rows, putting one over the other (partly overlapping). After about 2 days, palmyrah leaves with the petiole can be used for fencing. If it is for thatching, then they are usually arranged in a circle in several layers and finally heavy weights are kept over the heap of leaves to flatten the fronds. It is usually kept in this position for about 2 weeks. Some cut the petioles while thatching and others do thatching with the stalks. A more compact fence is made by closely arranging the leaf petioles vertically. In the Island of Timor (Indonesia), the petioles are arranged horizontally fitting one into the other in the form of a woven fence. In Jaffna, the partly stretched leaves (after remaining on the ground for 2 days) are used straight away for fencing. During fencing one leaf is placed at an angle, the next one at a spacing of 30 cm is placed vertically over it. This is repeated and the leaves are held in position by binding using petioles and tying with the 'naar' strips.

The first two tender unexpanded whitish leaves and the next 12 young green leaves are used for making various handicrafts. The whitish tender leaves are used for making soft fine handicrafts while the young green leaves are used for making stronger, but coarse textured utility items like mats, baskets, packaging material, inner lining of heavy duty fibre baskets, *etc.* In India, the mature leaf strips are woven into packaging materials to store and transport, dried fish, fresh fish, fruits, processed tamarind, yams and even ice. Experiments conducted at the Department of Botany, University of Jaffna have indicated that, more whitish leaves can be produced by covering some of the young leaves with a bag or cardboard material preventing light entering the leaves. Growing leaves covered in this manner for two or more months produced light coloured almost whitish (etiolated) leaves.

The usual practice is to remove only 2 tender leaves per year, but it this method is applied, it will be possible to obtain more such leaves. The tender leaves after sun drying are wrapped in gunny bags and store in a warm room to prevent discolouration and development of moulds. In India, the leaf products are fumigated over night by burning turmeric powder to retain the natural white colour. One of the problems with the leaf products is that, they discolour with time and become more brittle. Brittleness seems to increase with time due to loss of moisture and this could be prevented by giving some protective coating on the surface of the leaves. Treating the leaves with 5% sodium pentachlorophenate +0.1% lissapol (wetting agent) for about 20 minutes followed by 5% zinc sulphate for 5 minutes gave satisfactory results in preventing fungal attack, discolouration and brittleness. Their studies also showed that basic dyes like Malachite green (green colour), Bismarck brown (brown colour), Methylene blue (blue colour), Chlorazol orange (orange colour), Congo red (red colour) are good for dying leaves.

The palmyrah leaf based handicrafts are well developed in India. They produce a wide range of articles and they maintain high quality products. This is observed at the Manapad handicraft centre, managed by the Manapad Women Workers Palm leaf Industrial Cooperative Society Limited. This society is assisted by the Church of the area and has been sending regular in export orders to "OXFAM" in England and to a few other establishments in Scotland. Leaf based handicrafts are not well developed in Indonesia except in East Nusa Tengara region where handicraft items, such as hats, baskets, boxes *etc.* are being made. A musical instrument somewhat like a harp is made from the mature leaf in Timor.

In Sri Lanka, the traditional palmyrah leaf articles like mats, baskets, storage boxes of different sizes, strainers, winnow, trays to dry food material, are still popular and are used in most households particularly in the North and East. Quite a number of fishermen still transport fresh fish in large woven baskets made out of tender leaves. Dried fish and tobacco are still transported long distances in packages made out of palmyrah leaf. Only recently the PDB has started training centres to popularize the production of other handicraft items like hats, handbags, fruit trays, bottle covers and other fancy items. The average fresh weight of lamina of a single leaf is about 1.9 kg. On an average a palm has 40 leaves and the total weight of green matter available for cattle feed from a palm is around 76 kg. The mature leaf (lamina) is a good source of forage for cattle, especially in areas where there is scarcity of grass and other leafy materials. In some parts of Jaffna, the farmers depend solely on palmyrah to feed their cattle. About 4 leaves are enough to feed one cow per day. In this region palms are pruned regularly to obtain leaves for various purposes. Two methods of pruning are practiced on Palmyrah one is a hard pruning where most of the leaves are cut, leaving only 3-4 young leaves at the top of the crown. In other method, which is a moderate pruning where only about 12 of the oldest green leaves are cut. Hard pruning is done once in 2 or 3 years in regions where there is no tapping while moderate pruning is practiced in other places.

The leaf lamina, which is relished by cattle, provide a forage yield (6500 kg/ha), comparable to the finest forage legumes like alfalfa. When feeding, the lamina of the fresh leaf is cut from the petiole and all the midribs are removed leaving the strips of leaf blade. These are again split into smaller strips and shredded. At present no attempts have been made to convert the leaf into an animal feed, which can be stored.

Leaf petiole

The petiole of the palmyrah leaf, which is about 90 to 120 cm in length, is a fibrous material. The entire petiole is used as fencing material or as domestic fuel. It is also a source of fibre ('naar') where the yellow shiny skin on the inner flat surface of the petiole, from freshly cut leaves is peeled off. The peel with some of the inner spongy fibres still attached is used as a crude tying material. Usually the peel is sized into long narrow strips before using. In centres where the `naar' is processed for other uses, number of peels is bundled together and air- dried for 2 days. If not bundled, they bend upon drying. If it is necessary to store the fibre peels for some time, then they are covered with gunny bags to prevent discolouration and further drying. When the dried peels are used, they are soaked in water for 1/2 hour to 1 hour and the soft inner fibres are neatly removed using a sharp knife. Then the peel is sized to strips of long fibre of different width using sizing tools.. Wide strips of about 2 - 3 cm are used for making heavy duty loading and transporting baskets, while narrow strips of about 0.5 - 1 cm are used for weaving rattan, in making a number of utility items like shopping baskets, drivers seat rests, chairs, beds *etc.* The average weight of 100 dried "naar" strips is about 5.0kg. The shiny rather stiff posterior region of the petiole is also stripped and sized to provide edges for various baskets. The fibre of the spongy centre core of the leaf stalk is again used as tying material, commonly used for various agricultural operations like tying of animals, bundling of hay grass *etc*. Attempts were made to extract the fibres from the petiole with the view to use them like the fibres of the coconut husk. Yield of up to 15% obtained after retting the petioles in water for nearly 2 months. The tensile strength of the fibre was found to be much lower than coconut fibre and it is suggested that these fibres might be used only as filling material for cushions.

Leaf ribs ('Eekil')

Eekil is actually the fibrous ribs of the palmate segments of the frond. The number of such eekils per leaf varies from 60 to 80 and the length varies for 50 to 100 cm. The small ones are found at the two ends of the fan shaped leaf and are usually discarded. The long eekil is about 3 mm wide and 1.5 mm thick and 100 such long dried eekils weigh about 290 g. Palmyrah eekil is tough but flexible and can be woven to make various handicrafts. They are used predominantly for making baskets, winnows, and trays. Shorter lengths of the eekil are also used in brush making. These are also dyed red, brown or black.

a. The Mat industry

Mats made of palmyrah leaves are used by people for numerous purposes. The most popular among them would include:

- For drying food grains

- For roofing purposes

- For people to comfortably sleep and sit on

- For packaging purposes

The mats used for sitting and sleeping purposes are bit expensive. They are made with specially cut out and colored palmyra leaves. Such mats are reported to be extremely good for health.

b. Basket making industry

The baskets which are in everyday use produced out of palmyrah leaves are of different shapes and size. There are about 5 types of baskets. They are:

- Baskets used for carrying loads

- Baskets used for preserving eatables

- Baskets used for storing grains

- Baskets used for shopping purposes

- Baskets used for business purposes

In villages farmers and petty traders even today use the first type of baskets for carrying farm products such as coconuts, banana, tapioca and vegetable items. They are available in different shapes and size. Basket makers may charge as much as Rs.500/- for making an optimum size basket. The cost of a medium size basket is around Rs.250/-. Smaller baskets costs about Rs.100/-. Very small baskets used for collecting grains and vegetable items are sold for prices ranging from Rs.10 to Rs.15/-. These baskets are in very great demands.

c. Muram

Muram is otherwise known as sieves. It is used for separating grain from chaff or for removing dust and particles of stones or clay. Their popularity is on the vain due to the advent of readymade flour and quality tested and scientifically packaged with ISI marking rice, black gram green gram and wheat. But those who still depend on the conventional methods of cooking cannot do without these muram. The cost of the muram varies from Rs.150/- to Rs.200/.

d. Brush Making

Brushes used for toilet cleaning and floor cleaning are made out of palmyrah fiber. Palmyrah fiber is extracted out of the stalk of the palmyrah leaves. What is usually done is to cut off the bottom portion of these stalks and then they are subjected to a process of machine pounding for purposes of fiber extraction. These brushes are available in different size. The brush making industry in spite of the rising competition has been making remarkable progress over the years. Thus it is obvious that the palmyrah tree has been a source of the major inputs for numerous industries

Fibre Based Products

The basal sheath of the leaf called 'kanku mattai' in Tamil is a good source of long, stiff fibres. These fibres are actually vascular bundles or groups of sclerenchymatous fibre cells. The size and colour of the fibres vary; matured leaf sheaths have more of the dark brown or black strong fibres, while the younger leaf sheaths have relatively light coloured fibres.

Fibre is usually obtained from young palms of 5 to 10 years old. The leaves are cut in an orderly manner from the base upwards; cutting 12 leaves per year from a single palm. The sheath is cut carefully without damaging the stem with a sharp knife leaving a small stump of about 0.5 cm on the trunk. About 12 to 16 leaves from the apex downward are left on the palm. If leaves are harvested every year, the sheaths will yield only medium stiff (i.e. about 50% black fibres), but if harvested once in two years more prime stiff fibres (i.e. 80% black fibres).

Palmyrah fibre has certain desirable qualities besides being hard and stiff. They are resistant to both alkali and acids, resistant to water and attack by termites and other insects. It is for these reasons that palmyrah fibre is preferred to other vegetable fibres. They are used primarily for making sanitary industrial, domestic, scrubbing brushes and road sweeping brushes.

The sheath comes as a pair, is separated and each piece is about 25 - 60 cm long, 10 to 120 cm wide and 1.5 to 2.5 cm thick. Dried leaf sheath or those remain on the palm for many years, after the leaves have been cut and removed, are riot used for extracting fibre. In Sri Lanka, the sheaths are soaked in water and beaten to remove the fibre. But in India, leaf sheath (one day old) are beaten with wooden mallets, when fresh, to separate the fibre from the rest of the tissue. This practice saves time and the yield of fibre is also more. The beaten sheath is combed with the teeth of the petiole ('karukku' in Tamil). The fibre is then sized according to the length and bundled to about 10 cm diameter. This is known as 'Kora fibre'. One kilogram of processed fibre can be obtained from about 12 leaves (i.e. per palm). This is processed further, usually at the regional fibre processing centres. Here the bundles are loosened, combed on steel spikes, sized/picked and bundled according to length, given one tie, dried after spreading and re-stacked. The kora fibre contains a mixture of dark brown to black strong fibres and light brown to white soft fibres.

The bundles are then trimmed with a heavy knife on a wooden bed. Bundles of different length are made, 7-inch long (with 2 ties), 7 — 12 inch lengths (3 ties), and more than 12 inches (4 ties). The bundles are allowed to dry for sometime on cemented floor. The finished fibre bundles are usually baled into 50 kg weight bales with a dimension of 45 cm x 45 cm x 50 cm. in order to reduce the volume and save on freight charges. The baling machine used in India is operated manually and has a 20-ton pressing capacity.

Fibres are sometimes dyed and the colours used are black, ash, brown or red. Boiling the fibres in iron tanks, containing water mixed with powdered gallnuts, ferrous sulphate, lissopal, common salt, red/black dye and crude oil, dyes the fibres. Boiling is done for nearly five hours and thereafter the fibre is allowed to remain in the tank till next morning. The dyed fibre is then taken out of the tank and allowed to drain in chambers for 3 to 4 days. During this period, whitish fungi grow on the fibre. Then the fibre is spread out in the drying yard (cemented open floor) and dried partially. After this the fibre is first roughly bundled and after combing they are bundled neatly into hanks. As the finished hanks will contain some moisture they are spread in drying yard and well dried.

In Sri Lanka palmyrah fibre was extracted manually until 1991. The palmyrah development board of Sri Lanka initiated a project to mechanize some of the processes such as extraction of fibre and extraction of pulp from ripe fruits under the UNDP/DOSL (SRL/88/005) development project. A proto—type fibre extraction fibre machine consists of a rotary steel spiked comber mounted on to a 4 cm diameter shaft rotating at a speed of 450 - 500 rpm on double ball self aligning bearings was developed. The spikes on this machine are 5cm long, spaced at approximately 3 cm on a wooden cylinder of 30cm diameter and 45 cm in length.

Non- Edible Products From Palmyrah

Leaf Based Products

Gift pouches

Sweet boxes

Palmyrah leaf garland

 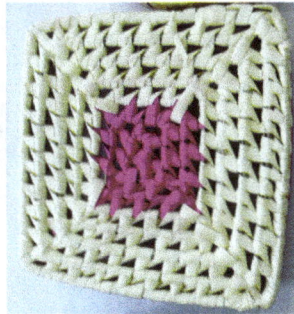

Pooja basket   Table mat

Platters   Basket and rings

Natural fiber woven dress made from palm leaves

Hat

Hand fan

Leaf woven mat

Toy

Bouquet

Palm leaf basketries

Fibre based products

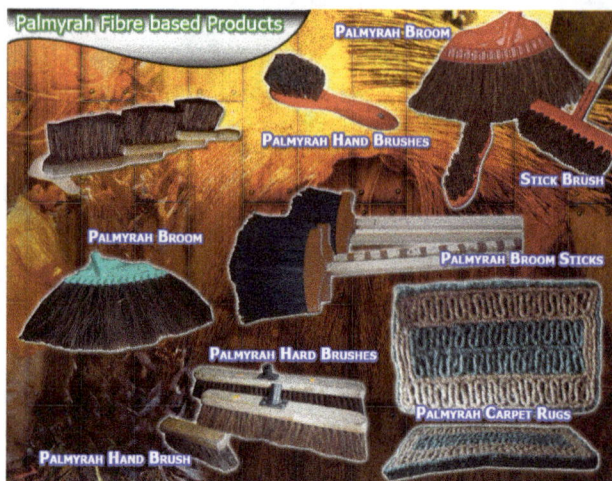

Coccothrinax crinita

Fibres of this palm are used for pillows, the trunk for shelter, and the leaves for bowls.

Cocos nucifera

Coconut has been used as food, as flavouring and made into beverages. The important products of coconut in our country are whole coconut (tender and mature), copra, toddy, neera, gur, coconut palm candy, sugar, vinegar *etc.* and few novel recipes like coconut *boli*, banana coconut cake. Moreover, it is used for making thatching, hats, baskets, furniture, mats, cordage, clothing, charcoal, brooms, fans, ornaments, musical instruments, shampoo, containers, implements and more

I. EDIBLE PRODUCTS

1. Neera and Toddy

Coconut toddy also called as palm wine is a sweetish, milk white liquid obtained when young coconut inflorescence is tapped. The unfermented sweet sap is called Neera which obtained before fermentation. A little lime is added to the collecting earthen pot to prevent the sap from fermenting. Toddy is obtained by natural fermentation of sap, when the sap flows from the spadix, fermentation starts suddenly. Flora such as bacteria and fungi are responsible for fermentation of coconut sap. Toddy is fully fermented in six to eight hours. Alcoholic content of toddy is about 4-6%

and had less shelf life. The main ingredient of the fermented sap is sucrose and there is very little reducing sugar, although other sugars like glucose, fructose, maltose and raffinose are present. Alcoholic liquid prepared from toddy called coconut arrack, which is generally distilled to between 33% and 50% alcohol by volume.

2. Snow ball tender nut

Snow ball tender nut is a round, soft, white coloured ball shaped tender coconut without husk, shell and testa. 8 months old coconut is more suitable for making snow ball tender nut. The tender coconut water can be consumed by inserting a straw through the top white tender coconut kernel. A groove is made in the shell by using a snow ball nut machine developed by Central Plantation Crops Research Institute (CPCRI), Kasargod, Kerala. By inserting a scooping tool, between the tender kernel and shell, the snow ball is scooped out from the shell. Since the tender coconut water retains its sterility since it is not exposed to the atmosphere. The shelf life can be prolonged to more than 10 days if the tender nut is individually packed and refrigerated under hygienic conditions.

3. Coconut oil

Coconut oil is extracted from the kernel or meat of matured coconuts harvested from the coconut palm. It contains more amount of saturated fat which leads to slow down oxidation resulting resistant to rancidity, lasting up to two years without spoiling. It has been reported that certain fatty acids and their derivatives of coconut oil can have detrimental effects on inactivating various microorganisms such as bacteria, yeast, fungi, and enveloped viruses. Coconut oil is generally obtained by mechanical extraction methods. In dry extraction method well dried copra is pressed in a screw press or hydraulic press to extract oil by breaking the oil cells in the kernel. After extraction, the oil and cake is separated by filtration.

4. Virgin coconut oil

Virgin coconut oil is said to be mother of all oils. It is rich in vitamin E, medium chain fatty acids particularly lauric acid, vitamins, minerals, antioxidant and exhibits good digestibility. About 50% lauric acid present in the virgin coconut oil which gives immune power to the human. The coconut oil produced through the wet method is known as virgin coconut oil. Virgin coconut oil was not subjected to high temperatures, solvents or refining process and therefore retains the fresh scent and taste of coconuts. The virgin coconut oil can be produced from fresh coconut meat or milk. It can be extracted from fresh meat by grating, drying and pressing. Producing it from coconut milk involves grating, mixing it with water and then centrifuging at high speed. The milk can also be fermented for 36–48 hours, the oil removed, and the cream is heated to remove any remaining oil. This method of extraction of coconut oil from coconut milk eliminates the use of solvent which may considerably reduce the investment cost and energy requirements. Moreover, it eliminates the RBD (refining, bleaching and deodorizing) process. Coconut oil helps to boost metabolism and raise body temperatures to promote thyroid health. It also heals the bruises, cures damaged tissues, moisturizes the skin, nourishes the brain, reverses the neurodegenerative diseases and prevents gastrointestinal mal absorption diseases.

5. Coconut protein powder

Coconut protein powder can be recovered from coconut wet processing waste which is obtained during the production of virgin coconut oil. The coconut milk from fresh and mature coconut undergoes protease treatment (100 tyrosine units/liter of coconut milk) for 2 hours in order to carry out effective destabilization of the coconut milk emulsion. Enzyme-treated milk is subjected to centrifugation at 7,000 rpm to obtain cream, coconut skim milk, and solid protein. Further, skim milk and solid protein is thoroughly mixed in the ratio of 8:2 v/w, homogenized and fed into a spray dryer. Then the protein powder is collected through a cyclone separator. The coconut protein powder showed high protein content of about 33 % and low fat content of 3 %. The protein powder had good emulsifying properties than skim milk protein and also had more water retention and swelling capacity than other dietary fibers.

6. Coconut honey

Coconut honey is viscous, free-flowing syrup, similar to coconut syrup but less creamy and less nutty in flavour was used as topping for pancakes and waffles. One part of skim milk was mixed with ½ part of refined sugar and ½ part of glucose, and then blended with sodium alginate at 0.5 per cent as stabilizer. Coconut cream may be added to improve the flavour of the product. The mixture was heated for 15 minutes, homogenized and cooked with constant stirring in steam-jacketed kettle to a TSS of 75 per cent. It was poured hot into sterile containers and then sealed hermetically.

7. Coconut sauce

A tangy sauce can be prepared from coconut water with red chilli, onion powder and little vinegar.

8. Coconut lemonade

It was prepared by boiling coconut water, sugar and lemon juice. It is traditionally a popular drink in Kerala.

9. Nata-de-coco

Nata from fresh liquid endosperm of the matured coconut was collected and filtered by using cheese cloth. It was pasteurized after adding 8 per cent sucrose, 0.5 per cent ammonium sulphate and adjusting the pH to 4.5 by adding acetic acid. The medium was developed by inoculating *Acetobacter acetii* at 10 per cent and incubated at room temperature for about two weeks. After two weeks a gel like mass developed with a film like layer on the top. The film was then removed and the mass was cleaned and cut into cubes. The cubes (nata) were further processed to improve the colour, flavour and taste by soaking it in sugar syrup. The nata can be used to decorate desserts, ice cream, puddings and fruit salad.

10. Coconut kernel or White meat

The preservation and pickling of coconut was done by soaking the coconut kernels in brine solution maintained their colour, flavour, texture and taste and resembled like fresh coconut for 90 days. The salt and acid penetration in pickle is directly proportional to the concentration in the soak solution.

The fresh coconut kernel could be preserved as long as for 3 to 6 months in the form of pieces and scrapings by steeping in a solution containing 4 per cent salt, acetic acid, sulphur dioxide and antioxidant. The product required washing in water prior to use.

11. Desiccated coconut

Desiccated coconut, the edible dried-out shredded coconut meat was prepared from fresh kernel of fully matured coconut and it is available in coarse, medium and fine grades and also in special grades such as threads, strips, granules *etc.* Good desiccated coconut is crisp, snow white in colour with a sweet, pleasant and fresh taste of coconut kernel. It is a commercial product was manufactured from the white part of the meat after removing the brown parings. The meat was shredded or disintegrated and dried in hot air driers at $140\text{-}170^0$F to 2 per cent moisture content and used in the manufacture of cakes, pastries and chocolates. It is the disintegrated, white kernel of coconut processed under strict hygienic conditions and dried to a moisture content of below 3.0 per cent. It is a food product which is ready and fit for direct human consumption.

Products from desiccated coconut

The matured coconut kernels were steam blanched and soaked in sugar syrup of 30^0 brix for a period of 48 hrs.The drained pieces were sulphited for 20 min and dried in the cabinet drier at a temperature of 60^0C for 8 h. Storage studies proved that sugar acts as an osmotic agent for the preservation of coconut. Osmotically dehydrated coconut was well suited for the preparation of products in the homes as well in commercial units.

The desiccated coconut powder (2%) or soy protein concentrate (SCP) (4.5%) was used in the preparation of mango bar, which increased the percentage of protein. The total soluble solids of pulp was raised to 30 brix with powdered cane sugar adding 0.6% citric acid and drying in an air cabinet drier at 63 ± 2^0C for 14h.

Coconut powder was used in the preparation of ragi based convenience mixes *viz.*, sweetened and spiced mix. The other ingredients for sweetened mix were puffed ragi flour, sugar, beaten rice flakes, cardamom, coconut powder and puffed ragi flour, sugar, dehydrated coriander leaves, hydrogenated fat, coconut powder and groundnut for spiced mix. In sweetened mix, addition of beaten rice improved the colour but increased the rate of lipid oxidation while incorporation of coconut powder lowered the lipid oxidation.

1. Coconut chips

Coconut chips, the thinly sliced crispy coconut meat which may be sweetened or salted and may come in handy as a ready -to- eat snack food. It was prepared by slicing the coconut meat of eleven to twelve month old nuts thinly into strands, soaked in syrup, drained and dried in a dryer or oven.

2. Coconut crisps

Coconut crisp is prepared from the young coconut endosperm of nine to ten month old nuts. It is white in colour, has pleasant coconut flavour and does not leave any fibrous feeling after taste. It was processed by slicing the meat into 0.6-0.7mm thickness, blanched in boiling water, cooked in light syrup and then dried which is considered as a high energy food and of a good quality product.

3. Roasted young coconut

Roasted young coconut was prepared by a process in order to sweeten its water and tender meat as well as to enhance their flavour. The process consisted of preliminary steps and the nuts were boiled in a solution (2% sodium meta bi- sulphite) for 20 min, dried and burnt for finishing touch and it was exposed to the fire for a minute or until the shell itself showed signs of burning. The fruit was kept at room temperature for 3 days or in the refrigerator for larger storage.

4. Honey roasted coconut

Honey roasted coconut, a sweet, thinly sliced crispy coconut meat, eaten as a snack food. Sliced matured coconut meats was mixed with honey sugar, margarine, sweeten condensed skim milk, molasses and salt and dried in oven for half an hour. Then it was agitated frequently for 15 min until it become crispy, cooled and packed.

5. Dehydrated sweet coconut

Young coconut meat was used for preparation of dehydrated sweet coconut. The coconut meat was washed with water and then soaked in coconut water, again washed with clean water. The meat was cut into pieces mixed with refined sugar and water in the ratio of 6:3:1 and boiled for 1 h or until it is dried, cooled and packed.

6. Dehydrated coconut chutney

Dehydrated coconut chutney using simple hot air drying technique was developed. It reconstitutes well in cold water and had all the characteristics of fresh chutney. The product had a shelf life of 3 months at 37^0 C and 6 months at ambient temperature when packed in flexible pouches.

7. Coconut milk

Coconut milk refers to the milky fluid, freshly extracted from the coconut kernel with or without added water and coconut cream to the high-fat cream-like material obtained from the coconut milk by either gravitational separation or centrifugation. It is prepared by blending skim milk powder with coconut milk obtained from freshly grated coconut and pasteurized at $70\text{-}72^0C$ for 10min.

Coconut cream, the concentrated milk extracted from fresh matured coconuts can either be used directly or diluted with water to make various curry preparation, sweets, desserts, puddings *etc.* Processed and packed coconut cream had a shelf life of six months and once opened it should be stored in refrigerator for subsequent use. It is a concentrated form of coconut milk, which is a convenient product prepared from mature and fresh.

8. Coconut syrup

Coconut syrup, a translucent, free-flowing liquid was prepared by cooking coconut milk with an equal amount of refined sugar and di-sodium-phosphate equivalent to 0.25 per cent of the volume of the milk, until the mixture attained a TSS content of

68-70 per cent. The hot mixture was poured in sterile containers and sealed hermetically. It was used as a topping for bakery products or as a mixer in alcoholic drinks or may be diluted in water and used in cooking rice cakes and other delicacies.

9. Sweetened condensed coconut milk

Coconut skim milk was used in the preparation of sweetened condensed coconut milk. Powdered dairy skim milk was added for protein fortification and other ingredients were corn oil, coconut cream and sugar. The skim milk was first pasteurized for 30 mins at $80-90^0$ C and mixed with other ingredients. The mixture was blended or passed through a colloid mill, and heated in a steam jacketed kettle with constant stirring to a TSS content of 68%. It was packed hot in sterile tin cans and cooled immediately in cooling tanks.

10 Coconut candy

Coconut candy was prepared from grated coconut meat mixed with coconut milk. The grated coconut was moistened with a portion of the milk. The remaining milk and the molasses were poured in a cooking pan, and the mixture was heated to boiling. Refined sugar was added and the mixture was cooked until it gets hardened when dropped into cold water. It was then poured in butter-greased pans, allowed to cool slightly, cut into desired sizes and individually wrapped in cellophane sheets.

11. Coconut jam

Jam is an intermediate moisture food prepared from the residual pulp left after removal of water from the kernels. Young tender coconuts are widely consumed as fresh can also be converted into value-added products such as jam. Coconut jam is prepared by boiling the pulp with sugar, pectin, acid, and other minor ingredients such as preservatives, coloring, and flavoring materials, to a reasonably thick consistency firm enough to hold the fruit tissues in position. The desired amount of sugar was added to the pulp mixture and heated continuously on under low flame. When the total soluble solids reached 60^0 brix, pectin (1.25 %) and citric acid (0.5 %) were added to the boiling pulp and the mixture was stirred continuously using a steel ladle. Heating can be stopped when the total soluble solids reached $67-68^0$ brix. The hot mixture was filled into sterilized glass bottles and cooled under ambient conditions. The prepared jam can be stored for a period of 6 months at ambient temperature without compromising the quality.

12. Coconut milk powder

Coconut milk powder was prepared by dehydrating the milk under controlled conditions. The composition of the milk was adjusted with fat percentage in the range of 50-60 per cent of the total solids. The emulsifiers and stabilizers were also added to the formulation. The most crucial step was the dehydration stage for which spray drier was employed at high temperature (around 180^0 C). Instant dehydration takes place converting each tiny droplet into microcapsules with fine droplet of the oil inside

13. Coconut flour

Coconut flour is a unique product prepared from coconut residue. Coconut flour, a by-product in the processing of coconut milk, can provide not only value-added income to the entrepreneurs but also a nutritious and a healthy source of dietary fiber for the consumers. It can be used as fillers, bulking agents and substitute for wheat flour, rice flour and potato flour at certain levels and incorporated into various food products like baked products, snack foods, steamed and extruded products. One by-product of coconut is the "sapal" from coconut meal, taken after extraction of the coconut milk. The "sapal" was made into coconut flour which contains dietary fibre. The utilization of coconut sapal may have some health benefits and may encourage the industry to produce value added products or functional foods which may help in the proper control and management of chronic diseases. This offers scope for utilization of coconut flour as a dietary component for diabetes. Low-fat, high fibre coconut flour, a unique product from sapal is a good source of dietary fiber. It is comparable with other cereal flour in terms of carbohydrate, fat and energy content and a good ingredient in nutraceuticals.

14. Tender coconut water concentrate

Coconut water concentrate was prepared using fresh coconut water collected under hygienic condition. Suspended solids and oil in the samples were removed by means of three way centrifuge. The removal of the solids and oil was necessary in order to minimize fouling of the membranes. The salts present in coconut water may be removed if desired, prior to concentration, to produce a very sweet product. This is achievable by passing the centrifuged coconut water through a mixed-bed ion exchange resin. The concentrate can be frozen or preserved in cans and used as base for the production of carbonated and non-carbonated coconut beverage.

Crystallized fruit was prepared using candied fruit coated with sugar or sugar crystals and allowed to deposit on it. Crystallized candy made with tender coconut kernel can be consumed as a snack. Tender coconut kernel in sugar syrup can be used as a dessert or incorporated in ice cream, custard or pudding.

15. Coconut jaggery

Coconut jaggery is prepared by the concentration of unfermented coconut sap. Jaggery is a wealthy source of calcium, iron and many other vitamins and minerals. It acts as a low calorie natural sweetening agent and digestive agent. The collected sap contains around 80% of water which has to be removed by evaporation. Before heating, the sap is filtered using sand filters to remove the impurities and a small quantity of alum is added to induce the precipitation of lime and magnesium. This will make the final jaggery much less deliquescent with a better colour and will remain hard for a reasonable period. After evaporation, a thick mass is obtained, which on further heating leads to crystallization and on cooling it sets a solid form. The final product has a dark colour due to the caramelization of sugar.

16. Coconut vinegar

Coconut vinegar, an alternate for synthetic vinegar, can be prepared from coconut water or from the sap of the coconut tree. It is extensively used as flavoring and preservative agents in pickles, salad dressings and sauces. It is rich in vitamins and minerals such as calcium, phosphorous, iron, sodium and also found to have anti inflammatory and anti microbial properties. Coconut vinegar is the resultant product of alcoholic and acetic fermentation of sugar enriched coconut water. After filtering, the coconut water is adjusted to 15^0 Brix by the addition of refined sugar and the mixture is heated to boiling point. The pasteurized mixture is then cooled and inoculated with active dry yeast (1.5 g/litre). After 5 to 7 days of alcoholic fermentation, mother vinegar or starter culture is added for effective acetic fermentation. The acidification process continued for up to 7 days and then the coconut vinegar is harvested by siphoning.

17. Coconut yoghurt

Yoghurt is a fermented product obtained by the fermentation of cow milk using lactic acid bacteria such as *Streptococcus thermophiles* and *Lactobacillus delbrueckii* spp. Bulgaricus. One liter of coconut milk was preheated at a temperature of 90^0C for 3 min. It was then cooled till the temperature reduced to 40^0C. 3% inoculum was mixed to the coconut milk and the cultured coconut milk was incubated at 37^0C for 8 hours then it was stored at 4^0C. A combination of soymilk (50%) and coconut milk (50%) has also been used in the preparation of soycoconut yoghurt.

18. Coconut refined sugar

The coconut sap is treated with two percent lime to coagulate albuminous impurities. The limed sap is then carbonated in two stages and filtered to remove surplus lime. The clarified liquid is evaporated to the extent of 75 per cent sugar content and the resultant syrup is concentrated in vacuum pans till crystallization begins. The syrup is then discharged into crystallizers and the crystalline sugar is separated by centrifugation. It is considered as low glycemic index food, that it will not raise the blood sugar levels. Low glycemic index food is good for the control of *diabetes mellitus* and it has been found that it lowers the bad cholesterol or low density lipoprotein cholesterol. Hence coconut sugar is considered to be healthier than refined cane sugar.

19. Coconut milk powder

This is a spray dried instant coconut milk powder, similar to dairy milk powder. It is packed in triple laminate pouches and has a shelf life of 12 to 24 months. The extraction of coconut milk for conversion into powder is the same as for coconut cream (milk). After adding certain ingredients, the coconut milk is concentrated and spray dried as in dairy industry.

20. Ready-to-use coconut chutney mix

Dehydrated coconut scrapings, roasted Bengal gram, ginger, garlic, green chillies, curry leaves, tamarind and salt were ground to get uniform slurry. The prepared slurry was seasoned in oil with mustard and curry leaves. The seasoned chutney was dried at 60⁰C for 6 h in a cabinet drier. For rehydration, 60 ml of water was added to the ready-to-use coconut chutney mix and allowed to stand for 2 minutes

21 Ready to use burfi mix

Ingredients

Dehydrated coconut scrapings	100g
Powdered sugar	150 g
Cardamom powder	2g

Method

- The dehydrated coconut scrapings are powdered coarsely using mixie. The powdered sugar and cardamom are added to the coarsely powdered coconut and mixed thoroughly.

22. Coconut cookies

Ingredients

Flour	60 g
Sugar	30 g
Shortening	60 g
Coconut powder	30 g
Cherry	25 g

Method

- Sieve the flour

- Place all the ingredients in a bowl except cherry and kneaded well

- Make small balls and placed it on a greased tray one inch apart.

- Place a piece of cherry above the ball.

- Bake at 275°F for about 12-20 minutes.

Edible Products From Coconut

Coconut milk powder

Coconut oil

Coconut candy

Coconut chips

Coconut toddy

Coconut vinegar

Coconut crisps

Coconut cookies

| Coconut kernel | Desiccated coconut |

II. NON-EDIBLE PRODUCTS

1. Value-added products from coconut shells

Like the coconut oil, the coconut shell also has exceptional properties. It has a specific gravity of 1.2, which is about twice the density of hardwood. It is at least twice as hard as hardwood and is also very rich in energy. The hardness of the coconut shell is comparable to lower strength aluminum alloys, making it one of the hardest organic materials produced in nature. It can be ground into 50-micron chips to potentially be used as reinforcement for engineering plastics. Chopped glass fibers are conventionally used as reinforcement to increase strength and stiffness and reduce cost in polymeric composites. Ground coconut shell is not as hard as glass, but it should bond much better to the matrix, since the bond interface will be organic to organic, rather than organic to silicon oxide. Because of its high mass-density, coconut shells also have a high energy-density. This means that they may be burned as fuel for cooking or used to make charcoal. While burning of 10 kg of wood produces only 1 kg of charcoal, 10 kg of coconut shells produce 3.5 kg of charcoal and 5.5 kg of combustible gases.

2. Value-added products from coconut husks

The husk is 35% of the mass of the coconut melon. It is comprised of about 67% pith, a lignin which behaves like a phenolic resin, and 33% fiber, also made from lignin but with a fibrous morphology. The pith and the fiber can be used in agricultural applications since they absorb about ten times their weight in water. Furthermore, the pith and fiber are biodegradable, enriching the soil much like peat or mulch, for which they may be substituted.

The husk can be hot pressed into particle board directly without adding any additional binder. The pith can apparently chemically react and consolidate much like a phenolic resin, with the fibers serving as reinforcement. Particle board in developing countries is usually in high demand and commands an excellent price. The particle board can also be used in the village for housing.

Coir is stiff coarse fiber that has been obtained from the outer husk of the coconut. The fibers range from sturdy strands suitable for brush bristles to filaments that can be spun into coarse, durable yarn. Coir has been traditionally used in the making of ropes and mats. Coir is a flexible natural yarn that is hauled out from mesocarp tissue, or husk of the coconut fruit. Generally coir is of rich yellow in color once it has been cleaned after the removal of the coconut husk; and hence it is often called " The Golden Fibre". Coir is the rubbery husk of the coconut shell. Being tough and naturally defiant to seawater, the coir guards the fruit sufficiently to endure months of floating on ocean currents to be washed up on a sandy shore where it may bud and mature into a tree, condition it has adequate fresh water, since all the extra nutrients it requires have been passed along with the seed. This uniqueness makes the thread fairly functional in floor and outdoor mats, aquarium filters, cordage and rope, and garden mulch.

Coir is used in brushes, doormats, mattresses, rope and twine, brooms and brushes, doormats, rugs, mattresses and other upholstery, often in the form of rubberized coir pads and sacking. A small amount is also made into twine. Pads of curled brown coir fibre, made by needle-felting (a machine technique that mats the fibres together) are shaped and cut to fill mattresses and for use in erosion control on river banks and hillsides. A major proportion of coir pads are sprayed with rubber latex which bonds the fibres together (rubberized coir) and is to be used as upholstery padding for the automobile industry. The material is also used for insulation and packaging. The major use of coir is in rope manufacture. Mats of woven coir fibre are made from the finer grades of bristle and white fibre using hand or mechanical looms. Coir is recommended as substitute for milled peat moss because it is free of bacteria and fungal spores. Hence, it is useful to deter snails from delicate plantings, and as a growing medium in intensive greenhouse cultivation.

a. Brown fibre

The fibrous husks are soaked in pits or in nets in a slow-moving body of water to swell and soften the fibres. The long bristle fibres are separated from the shorter mattress fibres underneath the skin of the nut, a process known as wet-milling. The mattress fibres are sifted to remove dirt and other rubbish, dried in the sun and packed into bales. Some mattress fibre is allowed to retain more moisture so it retains its elasticity for twisted fibre production. The coir fibre is elastic enough to twist without breaking and it holds a curl as though permanently waved. Twisting is done by simply making a rope of the hank of fibre and twisting it using a machine or by hand. The longer bristle fibre is washed in clean water and then dried before being tied into bundles or hanks. It may then be cleaned and 'hackled' by steel combs to straighten the fibres and remove any shorter fibre pieces. Coir bristle fibre can also be bleached and dyed to obtain hanks of different colours.

b. White fibre

The immature husks are suspended in a river or water-filled pit for up to ten months. During this time, micro-organisms break down the plant tissues surrounding the fibres to loosen them and the process is known as retting. Segments of the husk are then beaten by hand to separate out the long fibres which are subsequently dried and cleaned. Cleaned fibre is ready for spinning into yarn using a simple one-handed system or a spinning wheel.

c. Buffering

Because coir is high in sodium and potassium, it is treated before use as a growth medium for plants or fungi by soaking in a calcium buffering solution; most coir sold for growing purposes is pre-treated. Once any remaining salts have been leached out of the coir pith, it and the coir bark become suitable substrates for cultivating fungi. Coir is naturally rich in potassium, which can lead to magnesium deficiencies in soilless horticultural media.

Coir does provide a suitable substrate for horticultural use as a soilless potting medium. The material's high lignin content is longer lasting, holds more water, and does not shrink off the sides of the pot when dry allowing for easier rewetting. This light media has advantages and disadvantages that can be corrected with the addition of the proper amendment such as coarse sand for weight in interior plants like Draceana. Nutritive amendments should also be considered. Calcium and magnesium will be lacking in coir potting mixes, so a naturally good source of these nutrients is dolomitic lime which contains both. The addition of beneficial microbes to the coir media has been successful in tropical green house conditions and interior spaces as well. However, it is important to note that the microbes will engage in growth and reproduction under moist atmospheres producing fruiting bodies (mushrooms).

d. Bristle coir is the longest variety of coir fibre. It is manufactured from retted coconut husks through a process called defibring. The coir fibre thus extracted is then combed using steel combs to make the fibre clean and to remove short fibres. Bristle coir fibre is used as bristles in brushes for domestic and industrial applications.

Coir as technical textile

i. Use of coir as agrotech

Textiles used in Agriculture are termed as agro textiles. They are used for crop protection, fertilization. The essential properties required are strength, elongation, stiffness and biodegradation, resistance to sunlight and resistance to toxic environment. All these properties help the growth and harvesting of crops and other foodstuffs. There is a growing interest in using materials which gradually degrade.

Coir fibres when impregnated with natural rubber latex and moulded into suitable forms, can be used

For manufacturing various types of garden articles like supporting poles for climbers, pots for nursery plants, *etc.* Coir non-woven fabrics have been used for making hanging flower growing pots. Coir nettings have been utilized for establishing net houses which provide ideal growing place for delicate plants by cutting sunlight up to 50%. Such net houses could also find use for establishing roof gardening. Such nettings when applied as underlay in greenhouses have been found to enhance the growth and yield due to creation of adequate humidity and favourable conditions for the plants. Coir nettings when laid at a slope of around 60 degree, it can be used for growing vegetables like brinjal, ladies finger, *etc.* Some of the potential examples of use of coir as agrotech are

Preventing erosion and paving way for afforestation in greenhouse cover, preventing erosion and paving way for afforestation in greenhouse cover, for layer separation in fields, nets for plants, rootless plants and protecting grassy areas, as sun screens and wind shields, as packing material and in bags for storing grass/ tea leaves, controlling stretch in knitted nets, shade for basins, anti-bird nets, materials for ground and plant water management at the time of scarcity and abundance of water.

a. Plant Climbers or Coco poles (Gardening Coir Grow Stick)

Coco poles are made by wrapping coir twin around PVC pipes. Coco poles give ideal support to plants while creating a perfect moist environment for its roots. It's ideal for creepers like ornamental plants & even vegetables.

b. Coir baskets

Coir hanging baskets are made from coir fibre blended with natural rubber. Coir pots are 100% biodegradable used for the domestic gardening. Coir hanging baskets are used for internal gardening and acts as decorative materials to the buildings. Less weight and organic properties of coir makes them viable for internal gardening. Plants grow faster and healthier in coir baskets.

c. Coco pots (Moulded coir pots)

The moulded coir fibre pots are usually used as nursery bag for the seedlings, which can be directly planted without removing the 'nursery bag'. Coir fibre and natural rubber latex are used for the manufacture of coco pots; hence it is 100 % natural, biodegradable and eco friendly product.

d. Coir fibre discs (Tree cover)

It is made from 2 to 6 mm thick rubberized coir sheet and used for protecting the plant from evaporation. The fibre disc is also used as weed cover around the bottom of plant preventing direct sunlight thus avoids the growth of weed under the disc.

e. Coco chips (Husk chips)

The small fibre bits, which are a waste during the sieving of coir pith, can be mixed into pot plant media in horti-flori nurseries. All these materials find numerous applications in gardens and add a warm hue to the surroundings.

f. Coco peat

Coco peat is a by-product from coir defibreing units. This coco peat undergoes processing and is widely recognized as a superior growing medium for tomatoes, roses and many other crops. The Horticulture industry often calls this coco media as coco peat or coir peat.

g. Coir pith as briquettes

The coir pith is compressed into very small packs of 650 gm shrink wrapped which is most suitable to the hobby market and home garden owners.

h. Coco lawn

It is a readymade lawn and is an eco friendly method for faster development of readymade lawns using natural coir products instead of using synthetic lawns, which are costly, non-environment friendly and pose disposal problems. The coco lawn is a ready and easy to use eco-friendly alternative for various applications.

ii. Use of coir as buildtech

Buildtech textiles are used in construction – concrete reinforcement, facade foundation systems, interior construction, insulations, proofing materials, air conditioning, noise prevention, visual protection, protection against the sun and building safety.

Coir fibres when impregnated with phenol formaldehyde resin and converted into boards provide a wood substitute which finds use in the fabrication of low cost disaster management houses. This product named as coir wood finds use for manufacturing various articles like false ceiling boards, window and wall panels, furniture, corrugated roof sheets, *etc.*

Coir pith when pressed at certain temperature and pressure it can be converted into solid insulation boards which can be effectively used for keeping the buildings insulated from heat, cold and noise when laid on the inner walls.

a. Coir composite boards

Coir composites panels can be made using coir as reinforcing material with or without plantation timber (veneers like rubber, bamboo, jute, glass) in between as a secondary reinforcement and then infused with polymeric matrix material like phenolic, polyester, epoxy, *etc.*, and then processed under controlled temperature and pressure. Coir fibre composites has several advantages such as light weight, unbreakable, non-corrosive, water resistant, durable and affordable. Coir composites are highly suitable for building and construction for door, window panelling, furniture and other joinery work and transportation application for cost-effective replacement to wood and timber.

b. Acoustic barriers

Coir is being used as a noise prevention solution in homes located along highways and other high-traffic roads and also in offices. This is also used for making stylish compound walls and garden landscaping.

iii. Use of coir as clothtech

Technical textiles for clothing applications are especially in the finishing process where fabric is treated under pressure and high temperature, the technical textile supports the fabric for smooth processing.

Mobile coir fibre extraction machine developed by the coir board, extracts coir fibres and pith for the coconut husk in 10 seconds. The fibres obtained by this process are subjected to dew like spray treatment with a biochemical solution which bestows the coir fibres a soft and supple feel. Such fibres produce a soft and finer yarn of rummage up to 1000 metres/ kg which are being utilized for manufacturing cloths in union with cotton/ silk yarn to produce the garments like warm jackets for extremely cold climates at high altitudes. When these fabrics are treated under heat and pressure they will obtain smooth finish.

iv. Use of coir as geotech

Topsoil is essential to food production. Without it retention of water is greatly diminished. Additionally the sub layers beneath the topsoil are more compact and therefore water is more apt to runoff than to be absorbed. To make these exposed sub layers even marginally productive, large amounts of fertilizer need to be applied. Unfortunately, the available land for growing food is a finite nonrenewable resource. Moreover, we are losing topsoil faster than it is naturally produced (it takes 200- 1000 years to produce an inch of topsoil). Worldwide, the annual loss is about 26 billion tons. More and more, farmers have to be made aware about the beneficial use of coir geotextiles for protection from water and wind.

Coir geotextiles (Coir bhoovastra)

Coir geotextile is a permeable textile fabric in geotechnical engineering to prevent the soil from migrating while maintaining the water flow. Its role is to protect and promote vegetarian cover during its formative period after which it degrades over a period of time and mixes with the soil providing for valuable nutrients. It is resistant to rot, moulds, and moisture, and needs no chemical treatment. It is eco friendly and biodegradable. It is available in woven and non-woven forms.

It is made from coir fibre/yarn extracted from coconut husk either by natural retting or by mechanical process. It is a woven fabric of two treadle in construction made from coir yarn in which the warp and weft strands are positioned at a distance to get a mesh (net) effect of ¼, ½ and 1 inches. The netting (mesh) gives the grass plenty of room to grow, at the same time it provides large number of 'Check Dams' per square meter of soil surface. The nettings are normally produced on coir handlooms out of 2-ply coir yarn, with a width 1-2 meter and 50- metre length. It is estimated that India loses 27 per cent of top soil every year, losing 60 billion of tones of soil through erosion. Coir geo textiles are used for stabilization of soil through vegetation. These have been used to check the erosion of landscapes and soil slopes as well as protection of banks of river, canal and lakes, road and railway embankment. Reinforcement of mud walls of high velocity streams, bunds, farm and fishponds against erosion.

Compared to other natural fibres like cotton, jute, *etc*, coir fibres are of larger diameter and curvature. Coir fibres possess rigidity to bending which helps to bridge gaps in soil to perform as filters and for separation functions. Soil erosion is a great danger faced by the world with the rains washing away topsoil sans forest cover, vanishing at a very fast pace.

Coir geo textiles are also used for ground cover or mulch. As a ground cover, it reduces the flow velocity of runoff water by forming check dams with the help of net structured strands of coir geo textiles in firm contact with the soil, which absorb the impact of water flow and resist washing down keeping the soil intact. It provides support to the seeds sown and seedling, which could be otherwise easily washed away by water. The strands of the net reduce the wind velocity at the soil surface thereby trap soil particles from being blown away. As mulch, coir geo textiles provide ideal environment for the seeds to germinate and healthy growth of seedling by regulation of soil humidity, temperature and manure and controlling weeds, by protection from direct sunlight and rain.

Types of coir geotextiles

Open weave coir geotextiles

Open weave coir geotextiles are a net fabric woven from coir yarn. It is an ideal geo textile for situations for situations where land is sloppy which may lead to riling and gulling. In such slopes, heavy rainfall causes loss of soil. In the areas of scanty rainfall where soil is non- cohesive and prone to wind blowing, open weave coir geo textiles provides adequate protection. There is no need for post installation work.

The open weave coir geotextiles initially holds the ground for seeds and seedling an provides a mechanical support against water erosion, helps the germination of seeds for better and growth of the plants conserving moisture and adds organic matter to the soil after degradation. In areas where vegetation is poor or takes longer time for establishment, open weave coir geotextiles can hold the soil together for a longer period of time in comparison to other natural fibres.

Geo rolls and vegetation fascines

These are mainly used for the stabilization and revegetation of sites marked by steepness or high exposure to waves and currents causing instability. It is a construction modules characterized by a compact roll of coir web covered by exterior coir mesh netting making it strong and flexible. Their configuration and density help them to maintain form without losing material and promote plant growth as well as microbial activity. In areas where there is a constant flow of water, they facilitate new channel alignment. In standing water they initiate sedimentation, facilitate vegetation and dissipate induced wave energy. Geo rolls collect and hold mineral and organic particles, provide a physically stable substrate for root growth and gradually bio-degrade to leave a self sustaining erosion control system. The interior of the geo rolls consist of 100% coir fibre webs cross-lapped or air laid, followed by needle punching or stitch bonding. The fibre density is greater than or equal to 1000 gsm and the width of 220 mm to 600 mm. The substrate is then rolled into desired diameters.

Non-woven felts

In the manufacturing process, well cleaned coir fibres of good staple length pass through the cleaning machines by pneumatic suction and punched by the needle loom on one side to manufacture felts of different density depending upon punching intensity, needle penetration and thickness.

The fibre is mechanically bonded to form a continuous length of sheet. No bonding material is used in the manufacture. It can be manufactured in thickness from 10 mm to 20 mm with a density varying from 500 to 1500 g/sq.m. The felts have excellent moisture absorption and retention characteristics and form an ideal medium for plant growth.

Coir needled felts are available in blanket forms backed with nets made of jute, polypropylene and polythene also. The coir non-woven blankets are composed of 100% coir fibre randomly needle punched to the desired degree of compaction.

a. Coir blanket

Coir fibre stitched with PP or Jute between two netting either PP or Jute or a combination. Used to control soil erosion.

b. Cocologs

They are non-woven, made from coir fibre bunches under pressed condition in tubular enclosures of knotted coir yarn. Cocologs are mainly used for vulnerable streams, rivers or lake bank to protect scour. For high embankment areas with a variable water level, several cocolog can be applied as stack.

c. Coir fibre beds (Cocobeds)

Cocobeds are made from coir fibre and coir geo textiles. Coir fibre is sand witched between two coir geo textiles and stitched together to form a bed or pouch.

Steep stream banks can be covered with pre planted cocobeds. Sediments will be collected and held in cocobeds, which will help plant growth and purify water to a certain extent.

v. Use of coir as hometech

Textiles used in a domestic environment- interior decoration and furniture, carpeting, protection against the sun, cushion materials, fire proofing, floor and wall coverings, textile reinforced structures/ fittings fall in this category.

Coir mats and matting have been traditionally used for home furnishing. PVC and rubber tufted mats have found edge in the export market. With the development of fine yarn out of coir sisal blending it has been possible to produce coir curtains, venetian blinds and light weight mats. These products are also available in the dyed form using natural eco-friendly dyestuffs.

a. Mulch door mat

A traditional coir door mat beautifies the entrance to one's home or office and help one to keep the floors clean. It facilitate to keep the dirt, grime and moisture from coming indoor.

b. Mattress

This popular, natural fibre filled mattress is carefully made from coconut husk fibres which are bound together with natural latex. The coir pad is layered between foam and polyester wadding. It offers comfort, resilience and is available in any size or shape.

c. Coir fenders

Coir fenders are commonly made in spherical, cylindrical or ring shapes. Fenders are made in the required shape from coir yarn, rope or fibre tied together tightly. Coir rope knotted to form the outer shell by traditional knotting technique. Available in different diameters.

d. Coir mats

Coir mats are made on handlooms, power looms or frames and with or without brush. It is available in a range of colours, sizes and designs. The brushing qualities of coir doormats and their ability to keep the dirt away make the product a unique one. Mats are available in plain, natural and bleached, available with woven or stencilled designs and bevelled patterns for use in interior or exterior door fronts.

e. Coir matting

Coir matting is primarily used as a floor furnishing material. It is widely used in exhibition and fairs as a temporary but neat and elegant floor coverings. Because of its sound deadening characteristics, it is being used on a large scale for furnishing stairs, corridors, auditorium and cinema halls. A wide range of attractive designs and colours as well as quality makes it a favourite item for interior decorators.

f. Coir tiles

Mattings are cut, rubberised and finished with narrow straight edges, enabling it to be laid together to form tiles. Coir Tiles have inherent strength for durability and strength as it is made out of the strong fibres of coconut husks. Strong and eco-friendly coir tiles come in innovative designs so that it can make the floor look classy.

g. Coir rugs

Coir mattings cut to specified length, and suitably finished are marketed as 'Coir rugs'. Coir rugs are available in plain natural colour of the fibre, or in different shades, in woven patterns or printed designs. Rugs of various sizes with attractive designs are specially produced for overseas markets.

h. Coir mourzouk

A carpet in coir trade is called as mourzouk. Coir mourzouks are usually manufactured in a variety of sizes and patterns. They are mostly used for furnishing a selected area either at the centre of the room or any part of where generally the other portion might have furnished by other type of furnishing materials. They are also used for all-round furnishing.

i. Coir mattings for roof surface cooling

Cooling of buildings by roof surface evaporation is an established technology. It is an effective, simple, economical and environment friendly method of improving the indoor thermal conditions and reducing the capital and running cost of air-conditioning in the order of 60% and 30% respectively, under hot-dry conditions.

vi. Use of coir as indutech

Sound-proofing in industrial environment provides adequate scope of using coir fabrics. They are generally fire retardant; however, to enhance the properties for meeting international specifications, a formulation has been developed by coir board to make coir fabrics fire retardant. Railways have shown keen interest in the fire retardant coir for furnishing of railway coaches.

a. Coir belts

Coir belts are mainly used for driving machines and as conveyer belts.

vii. Use of coir as meditech

Coiryarn has been made finer and is blended with other natural fibres like sisal. The fine cloth is infused with the herbs of traditional Indian medicine as it has got the potential for curing many stress related conditions such as insomnia. It was named as Ayurcoir. The colours on the ayurcoir are produced from the medical preparation only and no other colourants are used. The roots, flowers, fruits, leaves, barks of different herbs are used to make dyes.

Coir fabrics have been used for creating healthy atmosphere for the patients suffering from diseases like arthritis, asthma, *etc.* The coir fabrics are dyed using eco-friendly natural colours and ayurvedic medicines which provide natural cure to the patients. It is gaining popularity due to its eco-friendliness. Nano cellulose has been recently extracted from coir pith which has potential uses in the wound dressings, *etc.*

viii. Use of coir as mobiltech

These textiles are used in the construction of automobiles, railways, ships, aircraft and spacecraft. Coir fibres when extruded with polypropylene could produce high tenacity products with low specific weight. Such products have potential use in the areas of making dash boards for automobiles and various parts in railways and ships, *etc.* Rubberised coir sheets provide high durability and cushioning properties by virtue of which they will find use in car seats. Non-woven coir felt has potential of use in the car trunk coverings with minimum airborne fibres unlike synthetics.

ix. Use of coir as oekotech (Ecotech)

Newer applications for textiles in environmental protection applications include floor sealing, erosion protection, air cleaning, prevention of water pollution, water cleaning, waste treatment/ recycling, depositing area construction, product extraction, domestic water sewage plants.

Cocolawn a readymade lawn developed by coir board using coir netting, coir non-woven, coir pith and coir pith organic manure has various eco friendly applications which include creating instantaneous greenery on rocky patches or any arid surface. The ready to use cocolawn is made available in the form of blanket, which can be shifted from one place to another and can be rolled upon for transportation. The harmful effects of chemical fertilizers to the environment are well known. Coir pith organic manure (C-POM) has been utilized as a source of soil less, pesticide free, nutritional medium for healthy growth of natural grass in an environment friendly manner. Replacement of soil makes the cocolawn lighter in weight also which helps in transportation and installation. Normally synthetic lawns are treated with ultraviolet radiation resistant chemicals to extend durability. The disposal of such synthetic lawns becomes a problem. The coir materials do not require such treatments as the presence of large amount of surface ligning nullifies the effect of UV light. Unlike synthetics, the cocolawn does not pose any ingestion risk to wild life. Typical applications include ground cover, roof cover and cycle path, footpaths and vegetation restoration in any denuded areas.

x. Use of coir as packtech

Coir wood has been used for making lids of packaging boxes for transportation of chemicals.

a. Coir yarn

Coir yarn is generally of 2 ply, spun from coir fibre by hand as well as with the help of traditional ratts, fully automatic spinning machines *etc*. Coir yarn is the raw material for the manufacture of a whole range of coir products. Today coir yarn is available in a range of colours that make them ideal for use in versatile applications.

b. Coir ropes

Coir rope-making is a common cottage industry in India. The Coconut fibre is attached to hooks on a wheel that is turned by hand. This twists the coir while more is added. It forms a strong rope that doesn't unwind or break. Among the natural fibre, coir has some unique characteristic particularly its rigidity, durability and friction.

The number of strands required for a strand of rope is determined by the diameter of the strand and the fineness of the yarn used. The diameter of the strand is in turn determined by the diameter of the rope and number of the strands constituting the rope. Therefore, by varying the number of yarns in the strand and the number of the strands, rope of any size can be made.

xi. Use of coir as protech

Coir umbrella developed by coir board is an effective Ultra Violet (UV) rays' cutter from sunlight providing protection to the human skin against UV rays. Such umbrellas could also find use for protection against sun rays on sea beaches, juice parlours, *etc*. Protective clothing *viz.,* a ballistic woven material for Body Armour could be developed using natural and synthetic kenaf and aramid in the ratio of 50:50.

xii. Use of coir as sporttech

Coir matting have been used as gymnasia matting and cricket pitch matting.

a. Coir mattings for cricket pitches

The coir mattings for cricket pitches are special type of matting. It may be provided with canvas or leather bindings at both ends. It is used to cover the 'Pitch' to protect it from the adverse effects of rain, moisture, storm and other natural factors. It is durable and requires low maintenance.

Non- Edible Products From Coconut

Coir As Agrotech

Coco pots (Moulded coir pots)

Coco bricks

Wall mounted coir baskets

**Coir fibre discs
(Tree cover)**

Coir fibre

Coir germination bag

Coco lawn

Coir vertical bag

Coir compost

Coir pith organic manure

Coir compost

Raw coir pith

Rubberized coir

Cocopeat grow bag

Coir net house

Coir based vertical garden

**Coco chips
(Husk chips)**

Coir As Buildtech

Coir wall panel

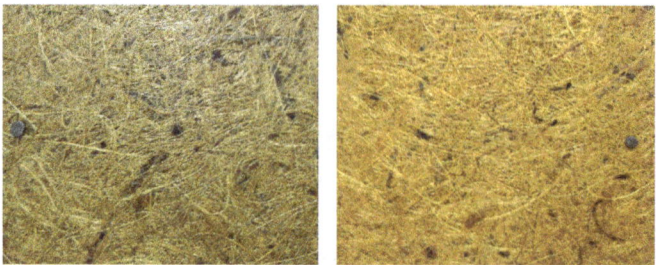

Coir composite boards

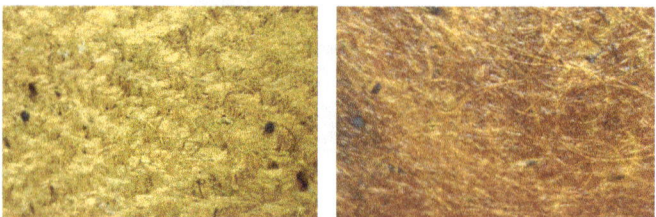

Coir composite boards

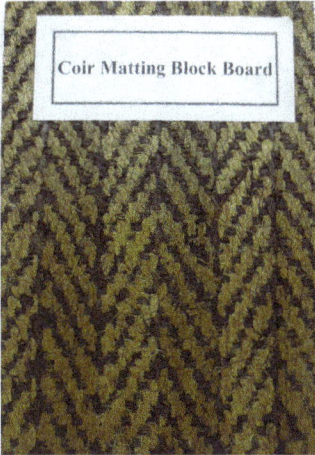

Coir matting black board

PVC tufted board

Coir polyester sheet

Coir furniture

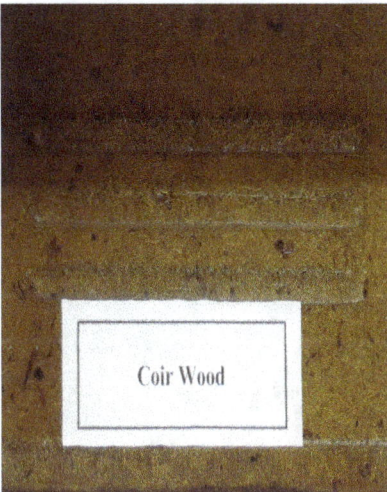

Coir wood

Wardrope made of coir based products

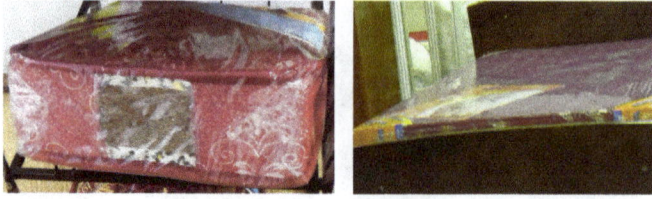

Coir mattress

Coir As Clothtech

Coir As Geotech

Coir geotextile

Coir roll

Coco log

Usage of coir geotextile, coir roll and coco logs on the bank of water canal

Coir As Hometech

Coir mats

Coir mats   Rubber tufted coir mats

Coir matting

Coir mourzouk

Coir mourzouk

Coir skin

Coir tiles   **Coir rug**

Coir tray

Coir wall hanging

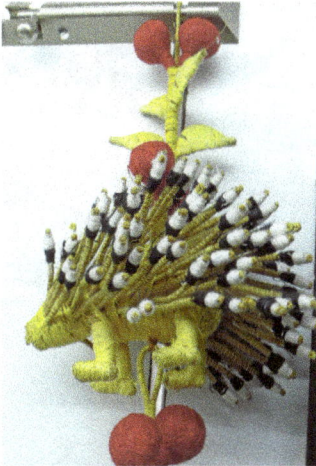

Coir wall hanging

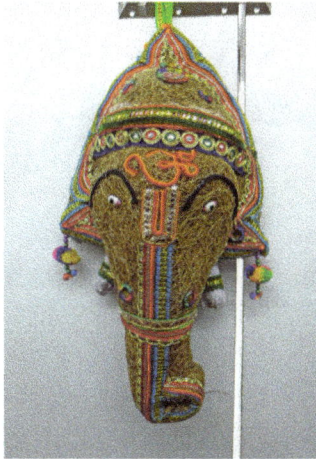

Coir wall hanging

Coir yoga mat

Coir chappals

Coir Handicrafts

Coir deer

Coir dinosaur

Coir dolphin

Coir ganapathy

Coir elephant

Coir elephant

Coir giraffe

Coir horse

Coir horse

Coir peacock

Boat in coir

Banana plant in coir

Coconut plant in coir

Coir lap top bag

Coir lady hand bag

Coir chair

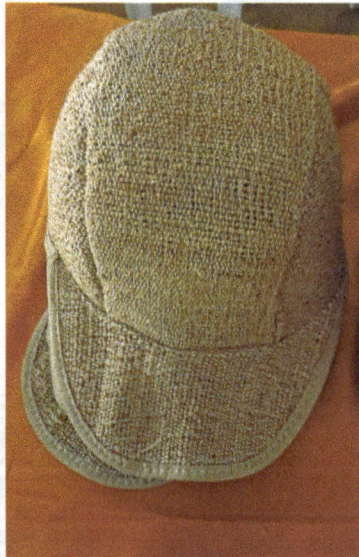

Coir cap

Taj Mahal in coir

Coir chess board

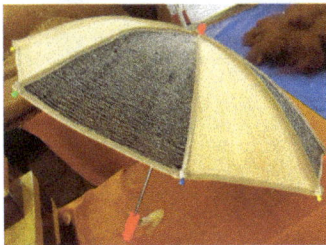

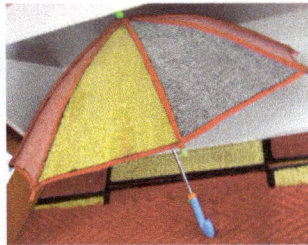

Coir umbrella

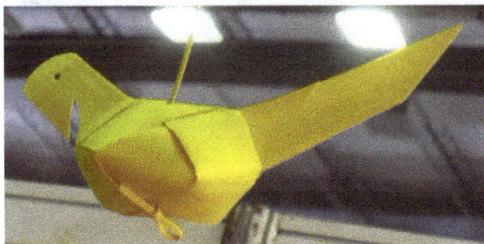

Coconut leaf bird

Coir bouquet

Coir bird's nest

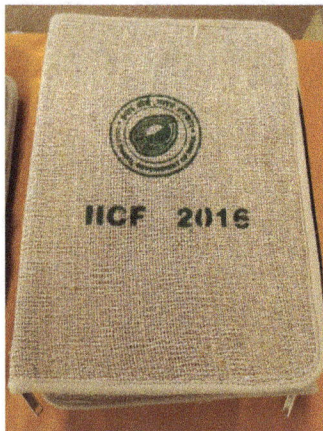

Coir file folder

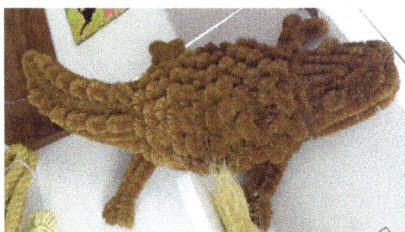

Coir crocodile

Coir fender

Coir Mouldings

Coir moulded products

Harvesting and
collection of coconut

Coir retting

Handloom weaving   **De-husking**

Coir Ornamentals

Fibre Based Products

Other Products From Coir

Bleached and dyed yarn

Coir ropes

Coir yarn

Bio oil and
ethanol from
coconut

Coir spools

***Copernicia baileyana*-** Leaves are used for weaving hats, baskets and also for thatch making

Corypha umbraculifera - The leaves are used for thatching and the sap is tapped to make palm wine

***Cyrtostachys renda*-** Stems are used for making flooring and leaves for thatch

Elaeis guineensis - Leaves are used for thatching; petioles and rachis for fencing

***Jubaea chilensis*-** The sap of this species has been used to make wine and also soft drinks

***Licuala orbicularis*-** The leaves are locally used for wrapping foods, manufacturing hats, umbrellas and other handicrafts

Metroxylon sagu - Fibre resulted after processing the pith and also from leaves may be used for mats. Dextrose sugar extract from sago palm starch can be processed to yield power ethanol

Oncosperma tigillarium - It is used for building construction, pig spears, palm heart edible

Phoenicophorium borsigianum - The large dried leaves are used for thatching

Phoenix sylvestris

- The trunk is used by the villagers in the construction of houses, it forming the supporting beam of the roof

- Halved trunks are used for diverting the water into the turbines of water-mills

- The leaves are used for making brooms, fans, floor mats, *etc*

- Flowers, borne on a spadix covered by a spathe which is 29.5 cm long; the spathe separates into two boat-shaped halves, exposing the flowers at maturity; both male and female inflorescences, about 25 cm long, bearing about 2,800 flowers

- The plants growing in the plains yield a good amount of juice which is used for making toddy and jaggery. The juice, as such, can also be drunk

- The tree provides a good fodder for milk cattle and is believed to increase the fat content of milk

- The fruit is cooling, oleaginous, cardiotonic, fattening, constipative, good in heart complaints, abdominal complaints, fevers, vomiting and loss of consciousness

- The juice obtained from the tree is considered to be a cooling beverage

- The roots are used to stop toothache

- The fruit pounded and mixed with almonds, quince seeds, pistachio nuts and sugar, form a restorative remedy

Rhopalostylis sapida - Leaves make a very effective waterproof roof and wall thatch and are woven into baskets and floor mats

Sabal palmetto

- Trunk is used for wharf pilings, docks, and poles.

- Brushes and whisk brooms are made from young leafstalk fibers

- Baskets and hats are made from the leaf blades

Syagrus romanzoffiana

- The leaves, or the fibers obtained from them, are used for making baskets and hats

- Its leaves and inflorescences are used as cattle fodder, especially for milking cows

- It is used for various constructional purposes, as stepping boards over swampy areas, footbridges and rustic piers in salt water

- The wood is moderately heavy, hard, very durable in salt water

Thrinax morrisii- Leaves are used for thatch making

Thrinax radiata - Leaves are used for roof thatching, broom construction, handicrafts, and food wrapping. Trunk have recently been used to construct lobster traps by fishermen

Annexure-List of palms

1. *Acoelorrhaphe wrightii* (Everglades palm, Paurotis palm, Saw cabbage palm, Silver saw palm, Silver saw palmetto)

2. *Adonidia merrillii* (Adonidia palm, Christmas palm, Dwarf royal palm, Manila palm)

3. *Aiphanes aculeata* (Chonta Ruro, Coyure Palm, Ruffle Palm, Spine Palm)

4. *Aiphanes erosa* (Macaw palm)

5. *Archontophoenix alexandrae* (Alex palm, Alexandra king palm, Alexander king palm, King palm, Northern bungalow palm)

6. *Archontophoenix cunninghamiana* (Bangalow palm, Piccabeen bungalow palm, Piccabean palm)

7. *Areca catechu* (Arecanut palm, Betel nut palm, Betel palm, Caccu, Catechu, Pinang)

8. *Arenga caudata* (Dwarf sugar palm)

9. *Arenga engleri* (Dwarf sugar palm, Sugar palm, Formosa palm)

10. *Arenga pinnata* (Aren, Areng palm, Black fibre palm, Gomuti palm, Kabong, Sugar palm)

11. *Arenga undulatifolia* (Aren Gelora)

12. *Bactris gasipaes* (Chonta, Peach palm, Pejibaya, Pejibeye, Pejivalle, Pewa, Pupunha)

13. *Bismarckia nobilis* (Bismarck palm)

14. *Borassus flabellifer* (Doub palm, Lontar plam, Palmyra palm, Pannaimaram, Tala palm, Talauriksha palm, Tal-gas, Toddy palm, Wine palm)

15. *Brahea armata* (Blue fan palm, Blue hesper palm, Grey goddess, Mexican blue fan palm, Mexican blue palm, Short blue hesper)

16. *Brahea edulis* (Guadaloupe palm, Guadalupe palm)

17. *Butia capitata* (Butia palm, Jelly palm, Pindo palm, South American jelly palm, Wine palm)

18. *Calamus australis* (Hairy mary, Lawyer cane, Lawyer's cane, Rattan palm, Wait-a-while, Wait-a-while palm, Wait-a-while vine)

19. *Carpentaria acuminata* (Carpentaria palm)

20. *Caryota mitis* (Burmese fishtail palm, Clustered fishtail palm, Fishtil palm and Tufted fishtail palm)

21. *Caryota urens* (Fishtail palm, Jaggery palm, Kitul tree, Sago palm, Solitary fishtail palm, Toddy palm, Wine palm)

22. *Ceroxylon quindiuense* (Andean wax palm, Wax palm)

23. *Chamaedorea cataractarum* (Cascade palm, Cat palm, Cataract palm)

24. *Chamaedorea elegans* (Good luck palm, Parlor palm, Parlour palm)

25. *Chamaedorea erumpens* (Bamboo palm)

26. *Chamaedorea metallica* (Metallic palm, Miniature fishtail plam)

27. *Chamaedorea microspadix* (Bamboo palm, Hardy bamboo palm)

28. *Chamaedorea seifrizii* (Bamboo palm, Reed palm)

29. *Chamaedorea stolonifera*

30. *Chamaerops humilis* (European fan palm, Fan palm, Mediterranean fan palm)

31. *Coccothrinax argentata* (Florida silver palm, Silver palm, Silver thatch palm, Silvertop, Silver top palm)

32. *Coccothrinax crinita* (Mat palm, Old man palm, Old man thatch palm, Palma petate, Thatch palm)

33. *Coccothrinax spissa* (Guano, Swollen silver thatch)

34. *Cocos nucifera* (Coco palm, Coconut palm, Coconut)

35. *Copernicia baileyana* (Bailey fan palm, Bailey's copernicia palm, Yarey, Yarey Hembra, Yareyon)

36. *Copernicia macroglossa* (Cuban petticoat palm, Jata de Guanbacoa, Petticoat palm)

37. *Corypha umbraculifera* (Talipot palm)

38. *Cyrtostachys renda* (Sealing wax palm, Lipstick palm, Maharajah palm)

39. *Dictyosperma album* (Common princess palm, Hurricane palm, Princess palm)

40. *Dypsis decaryi* (Three-sided palm, Triangle palm)

41. *Dypsis lutescens* (Areca palm, Butterfly palm, Cane palm, Golden feather palm, Golden cane palm, Madagascar palm, Yellow palm, Yellow butterfly palm)

42. *Elaeis guineensis* (African oil palm)

43. *Euterpe edulis* (Assai palm, Jucara palm)

44. *Hedyscepe canterburyana* (Big mountain palm, Umbrella palm)

45. *Howea belmoreana* (Belmore sentry palm, Curly palm, Sentry palm)

46. *Howea forsteriana* (Forster sentry palm, Kentia palm, Paradise palm, Sentry palm, Thatch leaf palm)

47. *Hydriastele wendlandiana* (Florence falls palm, Latrum palm)

48. *Hyophorbe lagenicaulis* (Bottle palm)

49. *Hyophorbe verschaffeltii* (Spindle palm)

50. *Hyphaene thebaica* (Doum palm, Branching palm, Ginger bread palm)

51. *Johannesteijsmannia altifrons* (Diamond Joey)

52. *Jubaea chilensis* (Chinese wine palm, Honey palm, Honey wine palm, Little cokernut, Syrup palm, Wine palm)

53. *Jubaeopsis caffra* (Pondoland palm)

54. *Laccospadix australasica* (Atherton palm, Queensland kentia)

55. *Latania loddigesii* (Blue latan, Blue latan palm)

56. *Latania lontaroides* (Red latan, Red latan palm)

57. *Latania verschaffeltii* (Yellow, Yellow latan palm)

58. *Licuala grandis* (Fan-leaved palm, Palas payung, Ruffled fan palm, Vanuatu fan palm)

59. *Licuala orbicularis*

60. *Licuala spinosa* (Mangrove fan palm, Spiny licuala)

61. *Linospadix monostachya* (Walking stick palm)

62. *Livistona australis* (Australian cabbage palm, Australian palm, Australian fan palm, Cabbage palm, Fan palm and Gippsland palm)

63. *Livistona chinensis* (Chinese fan palm, Chinese fountain palm, Fan palm, Footstool palm, Fountain palm)

64. *Lodoicea maldivica* (Coco-de-Mer, Double coconut, Seychellus nut)

65. *Lytocaryum weddellianum* (Dwarf coconut palm, Sago palm, Weddel palm)

66. *Metroxylon sagu* (Sago palm)

67. *Normanbya normanbyi* (Black palm)

68. *Oncosperma tigillarium* (Nibung palm)

69. *Parajubaea cocoides* (Coco Cumbe, Coquito, Mountain coconut, Quito palm)

70. *Phoenicophorium borsigianum* (Latanier palm)

71. *Phoenix canariensis* (Canary island date, Canary island date palm, Canary date palm)

72. *Phoenix dactylifera* (Date, Date palm)

73. *Phoenix reclinata* (African date palm, African wild date palm, Senegal date palm)

74. *Phoenix roebelenii* (Dwarf date palm, Miniature date palm, Pygmy date palm, Roebelin palm)

75. *Phoenix rupicola* (Cliff date palm, East India wine palm, India date palm, Wild date palm)

76. *Phoenix sylvestris* (India Date, Khajuri, Silver date palm, Sugar date palm, Sugar palm, Sugar palm of India, Toddy palm, Wild date, Wild date palm)

77. *Pinanga coronata* (Bunga, Ivory crown shaft palm, Pinang palm)

78. *Pritchardia pacifica* (Fan palm, Fiji fan palm, Pacific fan palm)

79. *Pseudophoenix sargentii* (Buccaneer palm, Cherry palm, Florida cherry palm)

80. *Ptychosperma elegans* (Alexander palm, Solitaire palm)

81. *Ptychosperma macarthurii* (Hurricane palm, Macarthur feather palm, Macarthur palm)

82. *Raphia farinifera* (Madagascar raphia palm, Latrum palm)

83. *Ravenea rivularis* (Majesty palm)

84. *Reinhardtia gracilis* (Window palm, Window pane palm)

85. *Rhapidophyllum hystrix* (Blue palmetto, Creeping palmetto, Dwarf saw palmetto, Hedgehog palm, Needle palm, Porcupine palm, Spine palm, Vegetable porcupine)

86. *Rhapis excelsa* (Bamboo palm, Fern rhapis, Ground rattan, Lady palm, Little lady palm, Slender lady palm, Miniature fan palm)

87. *Rhapis humilis* (Reed rhapis, Slender lady palm)

88. *Rhopalostylis sapida* (Brush palm, Feather duster palm, Nikau, Nikau palm, Shaving-brush palm)

89. *Roystonea regia* (Cuban royal , Cuban royal palm, Florida royal palm, Royal palm)

90. *Sabal minor* (Blue palmetto palm, Bush palmetto, Dwarf palm, Dwarf palmetto, Dwarf palmetto palm, Little blue stem, Scrub palmetto, Swamp palmetto)

91. *Sabal palmetto* (Blue palmetto, Cabbage palm, Cabbage palmetto, Cabbage tree, Common Palmetto, Palmetto, Palmetto palm)

92. *Serenoa repens* (Saw palmetto, Scrub palmetto)

93. *Syagrus romanzoffiana* (Giriba palm, Queen palm)

94. *Thrinax morrisii* (Brittle thatch, Brittle thatch palm, Broom palm, Buffalo-thatch, Buffalo top, Peaberry palm, Pimetta, Key palm, Key thatch, Silver thatch palm)

95. *Thrinax parviflora* (Broom palm, Florida thatch palm, Iron thatch, Mountain thatch palm, Jamaica thatch palm, Palmetto thatch, Thatch, Thatch pole)

96. *Thrinax radiata* (Florida thatch palm, Jamaica thatch, Sea thatch, Silk-top thatch)

97. *Trachycarpus fortunei* (Chinese windmill palm, Chusan palm, Fan palm, Hemp palm, Windmill palm)

98. *Trithrinax acanthacoma* (Buriti palm, Spiny fibre palm, Brazilian needle palm)

99. *Washingtonia filifera* (American cotton palm, Californian cotton palm, Californian fan palm, Californian palm, Cotton palm, Desert palm, Desert fan palm, Desert Washingtonia, Fan palm, Northern Washingtonia, Petticoat palm)

100. *Washingtonia robusta* (Mexican Washington palm, Mexican Washingtonia, Mexican fan palm, Skyduster Southern Washingtonia, Thread palm)

101. *Wodyetia bifurcata* (Foxtail palm)